METHODS IN MOLECULAR BIOLOGY

Series Editor
John M. Walker
School of Life and Medical Sciences
University of Hertfordshire
Hatfield, Hertfordshire, AL10 9AB, UK

For further volumes:
http://www.springer.com/series/7651

Hedgehog Signaling Protocols

Second Edition

Edited by

Natalia A. Riobo

Department of Biochemistry and Molecular Biology, Sidney Kimmel Cancer Center,
Thomas Jefferson University, Philadelphia, PA, USA

 Humana Press

Editor
Natalia A. Riobo
Department of Biochemistry and Molecular Biology
Sidney Kimmel Cancer Center
Thomas Jefferson University
Philadelphia, PA, USA

ISSN 1064-3745   ISSN 1940-6029   (electronic)
Methods in Molecular Biology
ISBN 978-1-4939-2771-5   ISBN 978-1-4939-2772-2   (eBook)
DOI 10.1007/978-1-4939-2772-2

Library of Congress Control Number: 2015942639

Springer New York Heidelberg Dordrecht London

Printed on acid-free paper

Humana Press is a brand of Springer
Springer Science+Business Media LLC New York is part of Springer Science+Business Media (www.springer.com)

Preface

Posttranslational Modifications and Secretion of the Hedgehog (Hh) Proteins

Mammalian genomes contain three homologs of the *Drosophila* Hh protein—Sonic (Shh), Indian (Ihh), and Desert (Dhh), encoded by separate genes [1]. All Hh proteins are auto-processed and posttranslationally modified in a similar manner before being secreted by the producing cell and received by surrounding tissues. Coincidentally, Hh proteins share a high homology in their C-terminal half, similar to the intein regions of self-splicing proteins of bacteria that undergo intramolecular processing. Autoprocessing occurs in the endoplasmic reticulum where the signal sequence of the Hh precursor is first cleaved. The C-terminal autoprocessing domain of Hh catalyzes the self-cleavage by an internal Cys nucleophilic attack on a peptide bond, which is resolved by addition of a cholesterol moiety that separates the Hh protein in two fragments, with the N-terminal fragment (HhN) covalently bound to cholesterol. The C-terminal fragment (HhC) is rapidly degraded by the proteasome [2], while HhN is further posttranslationally modified by addition of a palmitic acid group by skinny hedgehog (SKI) at its N-terminus before being targeted to the plasma membrane. Thus, active Hh proteins are dually lipidated, which makes them extremely hydrophobic. Release of HhN, from now on called simply Hh, to the extracellular milieu requires the activity of two proteins, a 12-pass transmembrane protein with homology to the Hh receptor called Dispatched (DispA in vertebrates), and the secreted glycoprotein Scube2, a member of the signal peptide, CUB, domain, epidermal growth factor-related family that is secreted together with Hh [3–5]. Scube2 interaction with Hh requires cholesterol modification of Hh, distinct from the interaction between Dispatched and Hh. The palmitic acid moiety enhances the Scube2-Hh interaction while also enhancing the rate of release from the producing cell. Although Scube2 is absent in *Drosophila*, alternative mechanisms appear to regulate its secretion. The lipid-binding protein Shifted (Shf), an ortholog of the Wnt Inhibitory Factor-1 (WIF-1) protein, works in a similar fashion as Scube2 by associating with cholesterol-modified HhN to promote its stability and release [6]. In addition, uptake of Hh proteins into lipoprotein-like particles promotes long-range signaling in *Drosophila*.

Lipidated Hh tethers to the plasma membrane with high affinity, but it must also signal over long distances in order to exert its morphogenetic role on cellular differentiation during embryonic development. Cholesterol-unmodified, full-length Hh is active, but not to the same extent as when lipidated. Moreover, signaling by unmodified Hh leads to ectopic expression of Hh target genes and ultimately developmental defects. A mechanism to explain the release of Hh from the producing cell must address the issues of solubility, stability, and activity of the protein once it is released. To address solubility, Hh may form oligomers that occlude the lipidated Hh in a "micelle-like" structure. A visible result of oligomerization is the formation of larger Hh-containing structure called visible clusters. In Chapter 1, a method for stable delivery of soluble Hh proteins in cell culture that can be adapted for studies in vivo will be described. Other carriers of soluble cholesterol-modified Hh have been identified and include lipoprotein particles and exosomes [5]. In addition, specialized filopodia known as cytonemes have been discovered as transporters of Hh, but not hubs for Hh signal transduction [7]. We will describe a detailed protocol for imaging of cytonemes in Drosophila wing discs in Chapter 2. To address stability, a subfamily of heparin sulfate proteoglycans (HSPGs) known as glypicans may be involved at least in *Drosophila* where the glypican Dally is involved in stabilizing Hh at the cell surface of Hh-producing cells. In vertebrates, the gene *tout-velu* (ttv) is required for the generation of HSPG chains and also provides a link between glypicans and Hh signaling [8]. In fact, glypicans play multiple roles in Hh signaling. For example, the glycosaminoglycan (GAG) chains of Dally and another *Drosophila* glypican, Dlp, mediate oligomerization or inclusion of lipid-modified Hh in lipoprotein particles in the *Drosophila* imaginal wing disc [9]. The signaling potency of Hh in vertebrates increases when lipidated. Conversely, lipidated Hh can only initiate juxtacrine signaling via direct cell-cell contact as it remains on the outer leaflet of the plasma membrane of the producing cell unless it is released into the extracellular environment. The distance over which the Hh signal is transmitted is intimately associated with Hh secretion. In Chapter 3 we present a mathematical model for studying Hh proteins gradient formation in tissues.

Hedgehog Reception and Signaling

In the absence of Hh, the 12-pass transmembrane receptor Patched1 (Ptch1) inhibits the activity of a G protein-coupled receptor (GPCR) and core transducer of the pathway, Smoothened (Smo), by keeping it in an inactive state via an unknown mechanism. The inhibitory role of Ptch1 is reversed by binding of an Hh ligand with the aid of at least one co-receptor. Three redundant co-receptors for Hh proteins have been identified: CAM-related/downregulated by oncogenes (Cdon), brother of Cdon (Boc), and growth arrest-specific 1 (Gas1). They are absolutely and collectively required for binding to Hh synergistically with Ptch1 to transmit the Hh signal, as Cdo$^{-/-}$;Boc$^{-/-}$;Gas1$^{-/-}$ triple knockout mice display phenotypes strongly associated with defective canonical Hh signaling and are unresponsive to Shh [10, 11]. Cdon and Boc are single-pass transmembrane proteins belonging to the immunoglobulin superfamily and share sequence homology with Interference Hedgehog (Ihog) and Brother of Ihog (Boi), respectively, in *Drosophila* [12, 13]. However, according to crystallographic and biochemical studies, their mode of Hh binding is completely unlike that of the vertebrate co-receptors [14]. Cdon and Boc share redundant functions in most cell types such as in cerebellar granule neuron precursor (CGNP) cells that require Boc and Gas1, but not Cdon, for proliferation [15]. Gas1 is a glycosyl phosphatidylinositol (GPI)-anchored cell surface protein. The co-receptors act upstream of Smo, as upregulation of Smo activity rescues Hh pathway-mediated proliferation in Boc$^{-/-}$; Gas1$^{-/-}$ CGNPs [15]. Glypicans are also GPI-anchored proteins that act as co-receptors for either promoting or inhibiting Hh signaling [8]. GPC1 and GPC3 are negative regulators of the Hh pathway, both interact with low-density lipoprotein receptor-related protein-1 (LRP1), and the interaction is stabilized in the presence of Hh. This stabilization induces endocytosis and degradation of the glypican–Hh complex. In addition, GPC3 competes with Ptch1 for Hh binding. GPC5 and GPC6 are positive regulators of the Hh pathway, as loss-of-function mutations in GPC6 contributes to autosomal recessive omodysplasia, and upregulation of GPC5 contributes to rhabdomyosarcoma, an Hh-dependent type of cancer [8].

Sequence similarity and crystallographic evidence support the notion that Smo belongs to the GPCR superfamily. The spatial organization of Smo's 7-pass transmembrane helical bundle resembles that of class A GPCRs; however, Smo lacks conserved residues in helix VI that are critical for facilitating the active state conformation in this GPCR subfamily [16, 17].

Interestingly, Smo shows high sequence homology to the Frizzled (FZD) family of receptors, members of class F GPCRs, in particular with the FZD cysteine-rich domain (CRD) that contains a hydrophobic groove to which cholesterol-like moieties bind [17]. Thus, Smo has been classified as a member of the class F GPCR subfamily [18]. *Drosophila* Smo also shares the GPCR-like heptahelical bundle; however the structures of vertebrate and invertebrate Smo proteins have diverged as they share low sequence homology [19].

Both *Drosophila* and vertebrate Smo couple to heterotrimeric G inhibitory proteins (Gi) [20, 21]. Evidence of Smo-Gi coupling in vertebrates derived from direct measurements of Smo-catalyzed exchange of GDP for GTP in all members of the Gi family [20]. These experiments showed constitutive activity of Smo toward Gi proteins in the absence of ligand and demonstrated that several Smo inhibitors act as inverse agonists preventing activation of Gi by constitutively active Smo [20]. A detailed description of the protocol to quantify the level of Smo coupling to G proteins is provided in Chapter 4. In addition, studies in cell culture using NIH3T3 cells, fibroblasts that respond readily to Hh proteins and that will be discussed in several chapters in this book, showed that pertussis toxin (PTX), a protein that disrupts coupling of GPCRs to Gi proteins, prevents many cellular responses to Shh and serves as a tool to study Smo-Gi-dependent processes. The most studied function of Gi proteins is inhibition of adenylyl cyclase and concomitant reduction in cytosolic cAMP levels. Accordingly, activation of Gi by Smo decreases the concentration of intracellular cAMP to a comparable extent than other prototypical Gi-coupled GPCRs [22]. The relevance of GPCRs and G proteins in Hh signaling continues to be more appreciated as recent evidence supports the role of the orphan GPCR Gpr161 in the cell-specific regulation of basal Hh signaling repression [23].

The canonical Hh pathway regulates the activity of the transcription factor Cubitus interruptus (Ci) in *Drosophila* and its three vertebrate orthologs glioma-associated oncogene homolog (Gli) transcription factors according to the degree of Smo activation [1]. A direct consequence of increasing Hh concentration is the stepwise phosphorylation of the Smo C-terminal tail (C-tail) by casein kinase 1 alpha (CK1α) and G protein-coupled receptor kinase 2 (GRK2) in vertebrates and by CK1α and protein kinase A (PKA) in *Drosophila* [24, 25]. A protocol for studying the phosphorylation events and its functional consequences is provided in Chapter 5. Phosphorylation of the C-tail of Smo leads to a conformational change that brings the C-tails of two Smo monomers together to form a functional dimer [26]. A method for analysis of Smo conformational changes is presented in Chapter 6. Activation of Smo correlates with its trafficking from intracellular vesicles to the plasma membrane in *Drosophila* and to the primary cilium in vertebrates [27, 28]. Smo translocation to the primary cilium requires the intraflagellar transport (IFT) protein complex component Kif3a, along with β-arrestins [29]. GRK2 phosphorylation creates docking sites for recruitment of β-arrestin 2 (βarr2) to active Smo at the plasma membrane [30]. Smo ciliary localization is not sufficient for activation of Gli, as it must undergo a second unknown activation step before it signals the activation of Gli at the primary cilium [31, 32]. Once activated by Smo, full-length activated Gli (Gli[A]) isoforms translocate to the nucleus where they activate the transcription of Hh target genes. Hh responsive genes are dependent either on Gli[A] activity or on the de-repression of Gli[R] activity [33]. The activation of Gli is intimately associated with the presence of the primary cilium, as full-length Gli must accumulate at the primary cilium along with the ciliary localization of active Smo in order for Gli activation to occur [34, 35]. The accumulation of Gli at primary cilia is PKA sensitive [34]. In Chapters 7 and 8, we present protocols for evaluation of transcriptional activity of the Gli transcription factors in cell culture and to determine the expression level of endogenous Gli proteins, which in the case of Gli1 serves as readout of canonical pathway activation, since it is the most sensitive Hh-target gene.

Ci activation in *Drosophila* requires the interaction between the C-terminal tail of the active Smo dimer and a Hh signaling complex (HSC) containing the kinase fused (Fu), the microtubule-binding protein costal2 (Cos2), suppressor of fused (Sufu), and Ci [36]. Interaction between Smo and the HSC results in: (1) autoactivation of Fu, (2) phosphorylation of Cos2, and (3) dissociation of Ci from the HSC. Autoactivation of Fu is required to antagonize the inhibitory activity of Sufu, whereas the dissociation of Ci from the HSC leads to its nuclear translocation by an unknown mechanism to activate Ci-target genes [37]. Thus, the relative stoichiometry among the HSC proteins is expected to govern Ci activation. A protocol for absolute quantification of different proteins of the Hh pathway is presented in Chapter 9.

In both *Drosophila* and vertebrates, in the absence of Hh ligands, Ci/Gli are kept in an inactive state by partial processing into transcriptional repressor forms, Ci^R/Gli^R. Phosphorylation by PKA at several sites initiates Gli^R formation at the base of the primary cilium, and Sufu forms a complex with Gli to prevent its translocation of Gli at the ciliary tip [38]. In Drosophila, Ci sequestration in the HSC allows PKA to initiate Ci^R formation. In order to signal partial proteolysis, PKA must phosphorylate Ci/Gli in several phosphorylation site clusters as complete loss of PKA catalytic activity is correlated with ligand-independent Hh pathway activation, and elevated intracellular cAMP levels are correlated with suppression of the Hh pathway [1, 33]. Phosphorylation of the first four out of six conserved Ser residues at the carboxyl side of the DNA-binding domain of Ci/Gli by PKA leads to further posttranslational modifications by glycogen synthase kinase-3 (GSK3) and casein kinase-1 (CK1) [39–41] and partial proteasomal processing. A protocol for determination of inhibitory and stimulatory phosphorylation of Gli2 is presented in Chapter 10, and a protocol for generating Gli2 and Gli3 with mutated phosphorylation sites expressed at near-endogenous levels is detailed in Chapter 11.

Full-length Ci/Gli in their active form are short-lived [33, 39]. In vertebrates, the Itch E3 ubiquitin ligase of the HECT family (also known as AIP4) ubiquitylates and targets Gli1 for total proteasomal degradation [42, 43]. The high promiscuity among E3 ligases and substrates predicts that in a context-specific manner, other E3 ligases might also regulate the turnover of the Ci/Gli transcription factors. In Chapter 12, we detail a protocol to investigate ubiquitylation of Gli family proteins by Itch that can be adapted to test other ligases and the type of ubiquitin chain branching. Another modification that inhibits Gli1 and Gli2 transcriptional activity in a reversible manner is acetylation by histone acetyltransferases (HATs) [44]. A method for evaluating the degree of Gli1 and Gli2 acetylation/deacetylation is detailed in Chapter 13.

Evaluation of Hh pathway activity in mammalian embryonic tissues has mostly been accomplished by in situ hybridization of Hh-target genes, such as *gli1* and *ptch1*, and by immunostaining of Hh ligands in those tissues that allow proper permeabilization maintaining the morphology. A novel method for evaluation of Ihh expression in the bone in chick embryos is presented in Chapter 14. In mice, the generation of a Ptc1$^{+/lacZ}$ heterozygote model served as reporter of canonical Hh signaling activity using β-gal staining of tissues as readout, since *ptc1* is a classical target gene. This was extremely useful to study Hh-responsive tissues during embryogenesis but has not been exploited as a common method to study Hh signaling in adult mouse tissues. Here we present a method for dual staining of β-gal and Shh in hematoxylin-eosin counterstained skin sections from Ptc1$^{+/lacZ}$ mice (Chapter 15).

Noncanonical Hh Signaling

Smo and Ptch1 are central components of the Hh pathway that leads to slow, graded activation of Gli (canonical signaling) but they can also initiate fast, nontranscriptional cellular responses collectively known as noncanonical Hh signaling [1, 45]. Overexpression of Ptch1 promotes cell death through its C-terminal domain and retains cyclin B1 out of the nucleus through its third intracellular loop, effectively slowing cell cycle progression [46, 47]. These functions of Ptch1 are independent of Smo and the Gli transcription factors and are classified as "Type I" noncanonical signaling [45]. The seventh intracellular domain of Ptch1, its C-terminal tail, recruits adaptor proteins and pro-caspase-9, which self-activates by oligomerization and induces apoptosis [48]. Importantly, Shh disrupts the Ptch1-proapoptotic complex interaction and promotes survival, as seen in endothelial cells [49].

Noncanonical Hh signaling Type II is mediated by Smo but is independent of Gli activation/repression. The ability of Smo to couple to Gi proteins is central to this branch of Hh signaling. It appears that Smo can activate Gi both within and outside of the primary cilium. For instance, Smo-Gi coupling in adipocytes and skeletal muscle stimulates aerobic glycolysis, glucose uptake, and calcium increase [50]. A protocol to study these metabolic effects of noncanonical Hh signaling is provided in Chapter 16. A similar primary-cilium localized activation of Gi by Smo regulates calcium spike activity in spinal neuron precursors [51]. In contrast, Smo-Gi coupling can occur outside of the primary cilium, as has been reported during activation of small Rho GTPases in fibroblasts and endothelial cells and of Src family kinases in neurons [52–54].

A comprehensive understanding of the Hh pathway and its functions in development and disease needs the consideration of both canonical and noncanonical responses. This Edition of Hedgehog Signaling Protocols provides novel protocols for the study of newly discovered functions and modifications of components of the Hh pathway.

Philadelphia, PA, USA

Lan Ho
Natalia A. Riobo

References

1. Robbins D, Feng D, Riobo NA (2012) The Hedgehog signaling network. Sci Signal 4 pt, 7
2. Chen X, Tukachinsky H, Huang CH, Jao C, Chu YR, Tang HY, et al. (2011) Processing and turnover of the Hedgehog protein in the endoplasmic reticulum. J Cell Biol 192: 825–838
3. Tukachinsky H, Kuzmickas RP, Jao CY, Liu J, Salic A (2012) Dispatched and scube mediate the efficient secretion of the cholesterol-modified hedgehog ligand. Cell Rep 2(2): 308–820
4. Creanga A, Glenn TD, Mann RK, Saunders AM, Talbot WS, Beachy PA (2012) Scube/You activity mediates release of dually lipid-modified Hedgehog signal in soluble form. Genes Dev 26(12): 1312–1325.
5. Briscoe J, Thérond PP (2013) The mechanisms of Hedgehog signalling and its roles in development and disease. Nat Rev Mol Cell Biol 14: 416–429
6. Glise B, Miller CA, Crozatier M, Halbisen MA, Wise S, Olson DJ, et al. (2005) Shifted, the Drosophila ortholog of Wnt inhibitory factor-1, controls the distribution and movement of Hedgehog. Dev Cell 8(2): 255–266
7. Gradilla AC, Guerrero I (2013) Hedgehog on the move: a precise spatial control of Hedgehog dispersion shapes the gradient. Curr Opin Genet Dev 23: 363–373
8. Filmus J, Capurro M (2014) The role of glypicans in Hedgehog signaling. Matrix Biol. 35: 248–252
9. Gallet A, Staccini-Lavenant L, Thérond PP (2008) Cellular trafficking of the glypican Dally-like is required for full-strength

Hedgehog signaling and Wingless transcytosis. Dev Cell 14(5): 712–725.

10. Allen BL, Song JY, Izzi L, Althaus IW, Kang JS, et al. (2011) Overlapping roles and collective requirement for the coreceptors GAS1, CDO, and BOC in SHH pathway function. Dev Cell 20: 775–787.

11. Izzi L, Lévesque M, Morin S, Laniel D, Wilkes BC, Mille F, et al. (2011) Boc and Gas1 each form distinct Shh receptor complexes with Ptch1 and are required for Shh-mediated cell proliferation. Dev Cell 20: 788–801

12. Yao S, Lum L, Beachy PA (2006) The ihog cell-surface proteins bind Hedgehog and mediate pathway activation. Cell 125: 343–357

13. McLellan JS, Yao S, Zheng X, Geisbrecht BV, Ghirlando R, Beachy PA, Leahy DJ (2007) Structure of a heparin-dependent complex of Hedgehog and Ihog. Proc Natl Acad Sci USA 103: 17208–17213

14. McLellan JS, Zheng X, Hauk G, Ghirlando O, Beachy PA, Leahy DJ (2008) The mode of Hedgehog binding to Ihog homologues is not conserved across different phyla. Nature 455: 979–983

15. Barzi M, Kostrz D, Menendez A, Pons S (2011) Sonic Hedgehog-induced proliferation requires specific Gα inhibitory proteins. J Biol Chem 286: 8067–8074

16. Wang C, Wu H, Katritch V, Han GW, Huang XP, Liu W, et al. (2013) Structure of the human smoothened receptor bound to an antitumour agent. Nature 497: 338–343

17. Nachtergaele S, Whalen DM, Mydock LK, Zhao Z, Malinauskas T, Krishnan K, et al. (2013) Structure and function of the smoothened extracellular domain in vertebrate Hedgehog signaling. Elife 2: e01340

18. Kristiansen K (2004) Molecular mechanisms of ligand binding, signaling, and regulation within the superfamily of G-protein-coupled receptors: molecular modeling and mutagenesis approaches to receptor structure and function. Pharmacol Ther 103(1): 21–80

19. Varjosalo M, Li S, Taipale J (2006) Divergence of hedgehog signal transduction mechanism between Drosophila and mammals. Dev Cell 10: 177–186

20. Riobo NA, Saucy B, DiLizio C, Manning, DR (2006) Activation of heterotrimeric G proteins by Smoothened. Proc Nat Acad Sci USA 103: 12607–12612.

21. Ogden SK, Fei DL, Schilling NS, Ahmed YF, Hwa J, Robbins DJ (2008) G protein Galphai functions immediately downstream of smoothened in Hedgehog signaling. Nature 456: 967–970.

22. Shen F, Cheng L, Douglas AE, Riobo NA, Manning DR (2013) Smoothened is a fully competent activator of the heterotrimeric G protein G(i). Mol Pharmacol 83: 691–697

23. Mukhopadhyay S, Wen X, Ratti N, Loktev A, Rangell L, Scales SJ, Jackson PK. (2013) The ciliary G-protein-coupled receptor Gpr161 negatively regulates the sonic hedgehog pathway via cAMP signaling. Cell 152: 210–223.

24. Chen Y, Sasai N, Ma G, Yue T, Jia J, Briscoe J, Jiang, J (2011) Sonic Hedgehog dependent phosphorylation by CK1α and GRK2 is required for ciliary accumulation and activation of smoothened. PLoS One 9: e1001083

25. Fan J, Liu Y, Jia J (2012) Hh-induced Smoothened conformational switch is mediated by differential phosphorylation at its C-terminal tail in a dose- and position-dependent manner. Dev Biol 366(2): 172–184

26. Zhao Y, Tong C, Jiang J (2007) Hedgehog regulates smoothened activity by inducing a conformational switch. Nature 450(7167): 252–258

27. Corbit KC, Aanstad P, Singla V, Norman AR, Stainier DY, Reiter JF (2005) Vertebrate smoothened functions at the primary cilium. Nature 437(7061): 1018–1021

28. Denef N, Neubüser D, Perez L, Cohen SM (2000) Hedgehog induces opposite changes in turnover and subcellular localization of patched and smoothened. Cell 102(4): 521–531

29. Kovacs JJ, Whalen EJ, Liu R, Xiao K, Kim J, Chen M, et al. (2008) Beta-arrestin-mediated localization of smoothened to the primary cilium. Science 320: 1777–17781

30. Chen W, Ren XR, Nelson CD, Barak LS, Chen JK, Beachy PA, et al. (2004) Activity-dependent internalization of smoothened mediated by beta-arrestin 2 and GRK2. Science 306: 2257–2260

31. Wilson CW, Chen MH, Chuang PT (2009) Smoothened adopts multiple active and inactive conformations capable of trafficking to the primary cilium. PLoS One 4: e5182

32. Rohatgi R, Milenkovic L, Corcoran RB, Scott MP (2009) Hedgehog signal transduction by smoothened: pharmacologic evidence for a 2-step activation process. Proc Natl Acad Sci USA 106: 3196–3201

33. Hui CC, Angers S (2011) Gli proteins in development and disease. Annu Rev Cell Dev Biol 27: 513–537

34. Kim J, Kato M, Beachy PA (2009) Gli2 trafficking links Hedgehog-dependent activation of smoothened in the primary cilium to transcriptional activation in the nucleus. Proc Natl Acad Sci USA 106: 21666–21671

35. Wen X, Lai CK, Evangelista M, Hongo J, de Sauvage FJ, Scales SJ (2010) Kinetics of hedgehog-dependent full-length Gli3 accumulation

in primary cilia and subsequent degradation. Mol Cell Biol 30: 1910–1922

36. Ogden SK, Ascano M Jr, Stegman MA, Suber LM, Hooper JE, Robbins DJ (2003) Identification of a functional interaction between the transmembrane protein Smoothened and the kinesin-related protein Costal2. Curr Biol 13(22): 1998–2003

37. Farzan SF, Stegman MA, Ogden SK, Ascano M Jr, Black KE, Tacchelly O, Robbins DJ (2009) A quantification of pathway components supports a novel model of Hedgehog signal transduction. J Biol Chem 284(42): 28874–28884

38. Tuson M, He M, Anderson KV (2011) Protein kinase A acts at the basal body of the primary cilium to prevent Gli2 activation and ventralization of the mouse neural tube. Development 138: 4921–4930

39. Wang B, Fallon JF, Beachy PA (2000) Hedgehog-regulated processing of Gli3 produces an anterior/posterior repressor gradient in the developing vertebrate limb. Cell 100(4): 423–434

40. Riobo NA, Lu K, Ai X, Haines GM, Emerson CP Jr. (2006) Phosphoinositide 3-kinase and Akt are essential for Sonic Hedgehog signaling. Proc Natl Acad Sci U S A 103(12): 4505–4510

41. Niewiadomski P, Kong JH, Ahrends R, Ma Y, Humke EW, Khan S, et al. (2014) Gli protein activity is controlled by multisite phosphorylation in vertebrate Hedgehog signaling. Cell Rep 6(1): 168–181

42. Di Marcotullio L, Ferretti E, Greco A, De Smaele E, Po A, Sico MA, et al. (2006) Numb is a suppressor of Hedgehog signalling and targets Gli1 for Itch-dependent ubiquitination. Nat Cell Biol 8(12): 1415–1423

43. Di Marcotullio L, Greco A, Mazzà D, Canettieri G, Pietrosanti L, et al. (2011) Numb activates the E3 ligase Itch to control Gli1 function through a novel degradation signal. Oncogene 30(1): 65–76

44. Coni S, Antonucci L, D'Amico D, Di Magno L, Infante P, De Smaele E, et al. (2013) Gli2 acetylation at lysine 757 regulates hedgehog-dependent transcriptional output by preventing its promoter occupancy. PLoS One 8(6): e65718

45. Brennan D, Chen X, Cheng L, Mahoney MG, Riobo NA (2012) Noncanonical Hedgehog signaling. Vitam Horm 88: 55–72

46. Thibert C, Teillet MA, Lapointe F, Mazelin L, Le Douarin NM, Mehlen P (2003) Inhibition of neuroepithelial patched-induced apoptosis by sonic hedgehog. Science 301(5634): 843–846

47. Barnes EA, Kong M, Ollendorff V, Donoghue DJ (2001) Patched1 interacts with cyclin B1 to regulate cell cycle progression. EMBO J 20(9): 2214–2223

48. Mille F, Thibert C, Fombonne J, Rama N, Guix C, Hayashi H, et al. (2009) The patched dependence receptor triggers apoptosis through a DRAL-caspase-9 complex. Nat Cell Biol 11: 739–746

49. Chinchilla P, Xiao L, Kazanietz MG, Riobo NA (2010) Hedgehog proteins activate pro-angiogenic responses in endothelial cells through non-canonical signaling pathways. Cell Cycle 9(3): 570–579

50. Teperino R, Aberger F, Esterbauer H, Riobo N, Pospisilik JA (2014) Canonical and non-canonical Hedgehog signalling and the control of metabolism. Semin Cell Dev Biol 33: 81–92

51. Belgacem YH, Borodinsky LN (2011) Sonic hedgehog signaling is decoded by calcium spike activity in the developing spinal cord. Proc Natl Acad Sci USA 108(11): 4482–4487

52. Polizio AH, Chinchilla P, Chen X, Kim S, Manning DR, Riobo NA (2011) Heterotrimeric Gi proteins link Hedgehog signaling to activation of Rho small GTPases to promote fibroblast migration. J Biol Chem 286(22): 19589–19596

53. Polizio AH, Chinchilla P, Chen X, Manning DR, Riobo NA (2011) Sonic Hedgehog activates the GTPases Rac1 and RhoA in a Gli-independent manner through coupling of smoothened to Gi proteins. Sci Signal 4(200): pt7

54. Yam PT, Langlois SD, Morin S, Charron F (2009) Sonic hedgehog guides axons through a noncanonical, Src-family-kinase-dependent signaling pathway. Neuron 62(3): 349–362

Contents

Contributors

ROBERT AHRENDS • *Project Group Quantitative Systems Analysis, Leibniz-Institut für Analytische Wissenschaften, Dortmund, Germany; Department of Chemical and Systems Biology, Stanford University School of Medicine, Stanford, CA, USA*

ROMINA ALFONSI • *Department of Molecular Medicine, University La Sapienza, Rome, Italy*

MARCUS BISCHOFF • *University of St. Andrews, St. Andrews, Scotland, UK*

JOÃO FRANCISCO BOTELHO • *Department of Biology, Faculty of Sciences, University of Chile, Santiago, Chile*

DONNA M. BRENNAN-CRISPI • *Department of Biochemistry and Molecular Biology, Thomas Jefferson University, Philadelphia, PA, USA; Sidney Kimmel Cancer Center, Thomas Jefferson University, Philadelphia, PA, USA; Department of Dermatology and Cutaneous Biology, Thomas Jefferson University, Philadelphia, PA, USA*

JUAN CALVO • *Centre de Recerca Matemàtica, Edifici C, Campus de Bellatera, Bellaterra (Barcelona), Spain*

GIANLUCA CANETTIERI • *Department of Molecular Medicine, Sapienza University of Rome, Roma, Italy*

SONIA CONI • *Department of Molecular Medicine, Sapienza University of Rome, Rome, Italy*

LAURA DI MAGNO • *Department of Molecular Medicine, Sapienza University of Rome, Rome, Italy; Center for Life NanoScience @ SapienzaIstituto Italiano di Tecnologia, Rome, Italy*

LUCIA DI MARCOTULLIO • *Department of Molecular Medicine, Center for Life Nano Science@Sapienza, Istituto Italiano di Tecnologia, Istituto Pasteur, Fondazione Cenci-Bolognetti, University La Sapienza, Rome, Italy*

SHOHREH F. FARZAN • *Department of Epidemiology, Geisel School of Medicine at Dartmouth, Lebanon, NH, USA*

LIN FU • *State Key Laboratory of Cell Biology, Institute of Biochemistry and Cell Biology, Shanghai Institute of Biological Sciences, Chinese Academy of Sciences, Shanghai, People's Republic of China*

ISABEL GUERRERO • *Centro de Biología Molecular (CSIC-UAM), Universidad Autónoma de Madrid, Cantoblanco, Madrid, Spain*

LAN HO • *Department of Biochemistry and Molecular Biology, Thomas Jefferson University,, Philadelphia, PA, USA*

CARMEN IBÁÑEZ • *Centro de Biología Molecular (CSIC-UAM), Universidad Autónoma de Madrid, Madrid, Spain*

PAOLA INFANTE • *Center for Life NanoScience at Sapienza, Istituto Italiano di Tecnologia, Rome, Italy*

JIANHANG JIA • *Department of Molecular and Cellular Biochemistry, Markey Cancer Center, University of Kentucky, Lexington, KY, USA*

KAI JIANG • *Department of Molecular and Cellular Biochemistry, Markey Cancer Center, University of Kentucky, Lexington, KY, USA*

NOAH RAY JOHNSON • *Department of Bioengineering, University of Pittsburgh, Pittsburgh, PA, USA; McGowan Institute for Regenerative Medicine, Pittsburgh, PA, USA*

XIANGDONG LV • *State Key Laboratory of Cell Biology, Institute of Biochemistry and Cell Biology, Shanghai Institute of Biological Sciences, Chinese Academy of Sciences, Shanghai, People's Republic of China*

MỸ G. MAHONEY • *Department of Biochemistry and Molecular Biology, Thomas Jefferson University, Philadelphia, PA, USA; Sidney Kimmel Cancer Center, Thomas Jefferson University, Philadelphia, PA, USA; Department of Dermatology and Cutaneous Biology, Thomas Jefferson University, Philadelphia, PA, USA*

DAVID R. MANNING • *Department of Pharmacology, Perelman School of Medicine, University of Pennsylvania, Philadelphia, PA, USA*

PAWEL NIEWIADOMSKI • *Department of Cell Biology, Nencki Institute of Experimental Biology, Warszawa, Poland*

VERÓNICA PALMA A • *Department of Biology, Faculty of Sciences, University of Chile, Santiago, Chile; FONDAP Center for Genome Regulation, Santiago, Chile*

SILVIA PANDOLFI • *Laboratory of Tumor Cell Biology, Core Research Laboratory, Instituto Toscano Tumori (CRL-ITT), Florence, Italy*

JOHN ANDREW POSPISILIK • *Max-Planck Institute of Immunobiology and Epigenetics, Freiburg im Breisgau, Germany*

NATALIA A. RIOBO • *Department of Biochemistry and Molecular Biology, Sidney Kimmel Cancer Center, Thomas Jefferson University, Philadelphia, PA, USA*

DAVID J. ROBBINS • *Department of Surgery, Miller School of Medicine, University of Miami, Miami, FL, USA*

RAJAT ROHATGI • *Department of Medicine and Biochemistry, Stanford University School of Medicine, Stanford, CA, USA*

ÓSCAR SÁNCHEZ • *Departamento de Matemática Aplicada, Universidad de Granada, Granada, Spain*

IRENE SEIJO-BARANDIARÁN • *Centro de Biología Molecular "Severo Ochoa" (CSIC-UAM), Universidad Autónoma de Madrid, Cantoblanco, Madrid, Spain*

FENG SHEN • *Department of Pharmacology, Perelman School of Medicine, University of Pennsylvania, Pennsylvania, PA, USA*

DANIEL SMITH-PAREDES • *Department of Biology, Faculty of Sciences, University of Chile, Santiago, Chile*

JUAN SOLER • *Departamento de Matemática Aplicada, Universidad de Granada, Granada, Spain*

BARBARA STECCA • *Laboratory of Tumor Cell Biology, Core Research Laboratory, Istituto Toscano Tumori (CRL-ITT), Florence, Italy*

RAFFAELE TEPERINO • *Institute of Experimental Genetics, Helmholtz Zentrum München, German Research Center for Environmental Health, Germany*

MARY N. TERUEL • *Department of Chemical and Systems Biology, Stanford University School of Medicine, Stanford, CA, USA*

YADONG WANG • *Department of Bioengineering, University of Pittsburgh, Pittsburgh, PA, USA; McGowan Institute for Regenerative Medicine, Pittsburgh, PA, USA; Department of Chemical and Petroleum Engineering, University of Pittsburgh, Pittsburgh, PA, USA; Department of Surgery, University of Pittsburgh, Pittsburgh, PA, USA*

YUE XIONG • *State Key Laboratory of Cell Biology, Institute of Biochemistry and Cell Biology, Shanghai Institute of Biological Sciences, Chinese Academy of Sciences, Shanghai, People's Republic of China*

YUN ZHAO • *State Key Laboratory of Cell Biology, Institute of Biochemistry and Cell Biology, Shanghai Institute of Biological Sciences, Chinese Academy of Sciences, Shanghai, People's Republic of China*

Chapter 1

Controlled Delivery of Sonic Hedgehog with a Heparin-Based Coacervate

Noah Ray Johnson and Yadong Wang

Abstract

Here we describe the preparation of a delivery vehicle for controlled release of Sonic hedgehog (Shh). The vehicle contains a synthetic polycation and heparin which interact by polyvalent charge attraction and rapidly self-assemble into liquid coacervate droplets. The coacervate loads Shh with high efficiency, protects its bioactivity, and provides sustained and localized release at the site of application. Shh coacervate may be injected directly or coated onto a polymeric scaffold for tissue engineering approaches, as described here.

Key words Sonic hedgehog, Controlled release, Coacervate, Heparin, Drug delivery, Scaffold

1   Introduction

Sonic hedgehog (Shh) is a powerful signaling molecule involved in the developmental patterning of nearly every organ in the body [1]. Accordingly, its therapeutic potential has been highly investigated in the context of the ischemic heart [2], ischemic hind limb [3], skin wound healing [4], bone repair [5], and diabetic neuropathy [6]. However, morphogens such as Shh have extremely short half-lives in the body, on the order of minutes to hours [7]; these proteins are rapidly degraded by proteases and removed into the circulation. A delivery vehicle that can protect Shh from the harsh environment, maintain it at the site of application, and slowly release it over time could extend its bioactivity and enable lower dosages to be used.

We have developed a delivery vehicle which can provide sustained, localized release of Shh. Our system is comprised of a synthetic biodegradable polycation, poly(ethylene argininylaspartate diglyceride) (PEAD) (Fig. 1a), designed to imitate the strongly cationic domains of the fibroblast growth factor receptor (FGFR) [8, 9]. The delivery system also contains heparin, the most negatively charged natural biomolecule. Exploiting its strong

Natalia A. Riobo (ed.), *Hedgehog Signaling Protocols*, Methods in Molecular Biology, vol. 1322, DOI 10.1007/978-1-4939-2772-2_1, © Springer Science+Business Media New York 2015

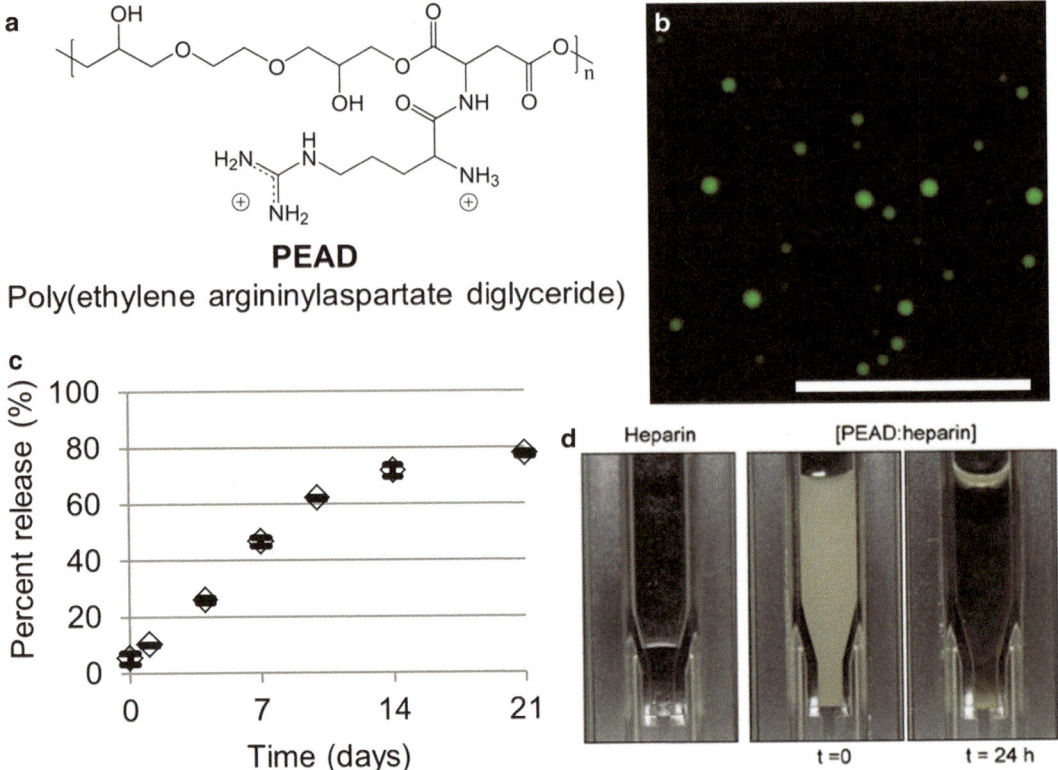

PEAD

Poly(ethylene argininylaspartate diglyceride)

Fig. 1 (**a**) Chemical structure of PEAD, the synthetic polycation, which interacts with heparin to form the coacervate delivery vehicle. (**b**) Fluorescent image of Shh coacervate showing spherical droplets with diameters on the nanometer-to-micron scale. Bar = 100 μm [11]. (**c**) Release profile of Shh from the coacervate over 3 weeks in vitro [11]. (**d**) Macroscopic view of heparin dissolved in saline as a clear solution (*left*), PEAD and heparin as a turbid solution immediately after coacervation (*middle*), and as a clear solution 24 h later as the coacervate settles to the bottom of the cuvette by gravity (*right*). Reprinted from Journal of Controlled Release, Vol 150, Hunghao Chu, Noah Ray Johnson, Neale Scott Mason, Yadong Wang, A [polycation:heparin] complex releases growth factors with enhanced bioactivity, Pages 157–163, 2011, with permission from Elsevier

heparin-binding affinity [10], Shh is completely loaded into the delivery system. PEAD is then added to neutralize the excess negative charges of heparin, which instantly induces self-assembly into micron-sized spherical droplets (Fig. 1b). The coacervate forms by charge-driven phase separation from the environment and appears as a turbid solution to the naked eye (Fig. 1d). The loading efficiency of Shh into the coacervate is greater than 95 %; release is slow and sustained for at least 3 weeks in vitro with no initial burst release in the first few days (Fig. 1c). The mechanism of Shh release is by hydrolytic degradation of PEAD [9]. We therefore anticipate that release will be accelerated in the body due to enzymatic activity that is absent in vitro.

We have reported the effect of Shh coacervate on cardiac fibroblasts and myocytes, demonstrating its potential as a regenerative therapy for cardiac ischemia [11]. Like Shh, numerous other therapeutic proteins including growth factors, cytokines, and morphogens also bind to heparin. Indeed, this coacervate serves as a platform technology for delivery of any heparin-binding protein, or co-delivery of any combination of heparin-binding proteins. We have validated its usefulness for delivery of basic fibroblast growth factor [12, 13], nerve growth factor [14], heparin-binding EGF-like growth factor [15], bone morphogenetic protein [16], stromal cell-derived factor-1α [17], vascular endothelial growth factor [18], hepatocyte growth factor [18], and bovine serum albumin [19], each with a distinct therapeutic application. In this chapter we focus on Shh; however, preparation of the coacervate for other heparin-binding factors is similar and may be widely utilized.

Here we describe in detail the steps to prepare the Shh coacervate and to perform a release assay to verify its release kinetics in vitro. We also explain the process used to coat a polymeric scaffold with the coacervate, endowing sustained release without clogging scaffold pores (Fig. 2a). The coating procedure produces a natural gradient which may be useful for stimulating cell infiltration into the scaffold ([17]; Fig. 2b, c). The coacervate could also be

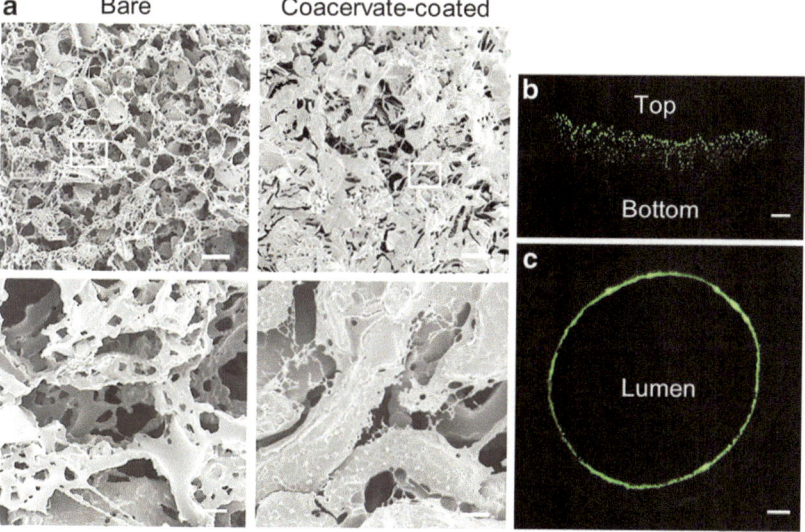

Fig. 2 (**a**) Scanning electron micrographs of the top surface of flat porous poly(glycerol sebacate) (PGS) scaffolds bare (*left column*) or coated with coacervate (*right column*) by applying by pipet to the top surface and then air drying. High magnification images (*bottom row*) are of the indicated fields in the *top row*. Bars = 100 μm (*top row*) and 10 μm (*bottom row*). (**b** and **c**) Cross section of flat and tubular porous PGS scaffolds coated with fluorescence-labeled coacervate by pipetting onto top surface (flat) or into the lumen (tubular) and then allowing to air-dry. The coacervate quickly absorbs into scaffold to form a natural gradient. Bars = 500 μm. Reprinted from Biomaterials, Vol 34, Keewon Lee, Noah Ray Johnson, Jin Gao, Yadong Wang, Human progenitor cell recruitment via SDF-1α coacervate-laden PGS vascular grafts, Pages 9877–9885, 2013, with permission from Elsevier

used to release anti-inflammatory factors from scaffolds to inhibit their rapid degradation following implantation. If a tissue engineering scaffold is not applicable, the coacervate may also be injected into tissues through as small as a 34 gauge needle, as the droplets are 0.5–10 μm in diameter (Fig. 1b).

2   Materials

1. 10 mg/ml heparin dissolved in saline (Heparin Sodium USP (Scientific Protein Laboratories LLC, Waunakee, WI)). Filter-sterilized at 0.2 μm.

2. Sonic hedgehog protein: Recombinant mouse N-terminus peptide (R&D Systems, Minneapolis, MN).

3. 10 mg/ml PEAD (poly(ethylene argininylaspartate diglyceride)), synthesized as described previously [9], dissolved in saline. Filter-sterilized at 0.2 μm.

4. Saline: 0.9 % NaCl injection USP (Baxter, Deerfield, IL).

5. Syringes.

6. 0.2 μm syringe filters.

7. Homemade or commercially available polymeric scaffolds of desired shape and size.

3   Methods

3.1 Shh Protein Reconstitution

1. Reconstitute Shh protein in saline to a stock concentration of 0.1 mg/ml (*see* **Note 1**).

2. Aliquot and store at −20 °C (*see* **Note 2**).

3.2 Preparation of Shh Coacervate

1. Thaw Shh protein stock solution and further dilute to 0.01 mg/ml in saline (*see* **Note 3**).

2. Combine Shh protein and heparin solution first at a 1:1 volume ratio of Shh–heparin (100:1 mass ratio) and mix briefly by pipetting.

3. Add PEAD solution at a 5:1 volume ratio of PEAD–heparin (5:1 mass ratio) and mix briefly by pipetting. The solution should become turbid immediately following the addition of PEAD (*see* **Notes 4–6**).

4. Use the coacervate immediately if the desired formulation is submicron droplets. Otherwise, the coacervate will precipitate to form an oily phase at the bottom of the container. The Shh bioactivity is protected in either the submicron droplet form or the bulk oil form. If needed, the suspension of submicron droplets may be reformed by pipetting up and down the precipitated "oil" within water.

3.3 Shh Release Assay

1. Prepare coacervate containing 100 ng Shh in a centrifuge tube (*see* **Note 4**). Then add 130 µl more saline for a total volume of 200 µl.

2. Centrifuge for 5 min at $12,100 \times g$ to pull the coacervate down and form a small white pellet near the bottom of the tube.

3. Aspirate the entire supernatant using a pipet, taking care not to touch or disrupt the pellet (*see* **Note 7**).

4. Store the supernatant (release fraction 1) at –20 °C until further analysis.

5. Replace supernatant with 200 µl fresh saline and store tube at 37 °C to simulate in vivo conditions.

6. Repeat collection procedure (take the entire supernatant as release fractions) at each time point desired, adding fresh saline (200 µl) each time (*see* **Note 8**).

7. After all time points have been collected, thaw all release fractions and analyze Shh concentration by any method preferred, e.g., ELISA or western blot (*see* **Note 9**).

3.4 Coating a Scaffold with Shh Coacervate

We have coated porous poly(lactic-*co*-glycolic acid) (PLGA) and poly(glycerol sebacate) (PGS) scaffolds synthesized in our lab by the salt-leaching method. In theory, any "homemade" or commercially available polymeric scaffold may be coated by the techniques described here. Shh coacervate should first be prepared as described in Subheading 3.2. Following scaffold coating, morphology may be observed by scanning electron microscopy (SEM) after thorough drying and sputter coating. Once coated, scaffolds will retain bioactivity when stored at 4 °C for at least 1 month. For long-term storage beyond 1 month it is preferred to lyophilize the scaffolds and store them at –20 or –80 °C.

3.4.1 Flat Hydrophobic Scaffolds

1. Lay scaffold flat and apply Shh coacervate directly to the top surface by pipet.

2. Allow scaffold to air-dry in a clean Biosafety Cabinet (*see* **Note 10**).

3.4.2 Flat Hydrophilic Scaffolds

1. Hydrophilic scaffolds may be dipped into Shh coacervate solution, which will be rapidly absorbed, or Shh coacervate may be applied preferentially by pipet (*see* **Note 11**).

2. Allow scaffold to air-dry in a clean Biosafety Cabinet (*see* **Note 11**).

3.4.3 Tubular Porous Scaffolds

1. Sit tubular scaffolds upright, on end, on a hydrophobic surface (e.g., Teflon) and fill the lumen with an appropriate volume of Shh coacervate by pipet (*see* **Note 10**).

2. Allow scaffold to air-dry in a clean Biosafety Cabinet (*see* **Note 11**).

4   Notes

1. Protein should be resuspended carefully by pipet to avoid frothing.

2. Reconstituted protein is stable for 3 months at –20 to –70 °C.

3. Once thawed, Shh protein solution can be stored at 4 °C for up to 1 month. Do not refreeze as this will significantly reduce protein bioactivity.

4. For example, to form coacervate containing a total dose of 100 ng Shh, first combine 10 μl heparin and 10 μl Shh protein solution, then add 50 μl PEAD.

5. Stock concentrations and volumes may be adjusted as desired to fit a particular application; however an overall mass ratio of 1:100:500 of Shh–heparin–PEAD should be maintained. If stock concentrations are reduced significantly then coacervate solution turbidity will be less apparent.

6. PEAD and heparin solutions are stable at 4 °C for at least 1 month.

7. Based on the centrifuge rotor angle the pellet should form on one side of the tube near the bottom. When aspirating, begin with the pipet tip just below the surface of the liquid and then move slowly down wall of the tube opposite the pellet as you aspirate.

8. The first sample collected immediately after forming the coacervate is day 0 release, or effectively the loading efficiency. Repeat, for example, on day 1, 4, 7, 10, 14, and 21. If reporting standard deviations is desired, release assay may be performed in multiple separate tubes at the same time.

9. We have found the Mouse Shh DuoSet ELISA Development Kit (R&D Systems #DY461) to work well following the manufacturer's instructions.

10. Shh coacervate solution will absorb into scaffold pores by capillary action (Fig. 2a); coacervate droplets will preferentially adsorb to the scaffold, thereby forming a gradient (Fig. 2b, c).

11. Drying time will vary based on volume applied. To accelerate drying, scaffolds may also be lyophilized in a bottle assembled and then de-assembled within the Biosafety Cabinet.

References

1. Ingham PW, McMahon AP (2001) Hedgehog signaling in animal development: paradigms and principles. Genes Dev 15:3059–3087

2. Kusano KF, Pola R, Murayama T, Curry C, Kawamoto A, Iwakura A et al (2005) Sonic hedgehog myocardial gene therapy: tissue repair through transient reconstitution of embryonic signaling. Nat Med 11:1197–1204

3. Pola R, Ling LE, Silver M, Corbley MJ, Kearney M, Blake Pepinski R et al (2001)

The morphogen Sonic hedgehog is an indirect angiogenic agent upregulating two families of angiogenic growth factors. Nat Med 7: 706–711

4. Asai J, Takenaka H, Kusano KF, Li M, Luedemann C, Curry C et al (2006) Topical sonic hedgehog gene therapy accelerates wound healing in diabetes by enhancing endothelial progenitor cell-mediated microvascular remodeling. Circulation 113:2413–2424

5. Rivron NC, Raiss CC, Liu J, Nandakumar A, Sticht C, Gretz N et al (2012) Sonic Hedgehog-activated engineered blood vessels enhance bone tissue formation. Proc Natl Acad Sci U S A 109:4413–4418

6. Calcutt NA, Allendoerfer KL, Mizisin AP, Middlemas A, Freshwater JD, Burgers M et al (2003) Therapeutic efficacy of sonic hedgehog protein in experimental diabetic neuropathy. J Clin Invest 111:507–514

7. Kicheva A, Bollenbach T, Wartlick O, Julicher F, Gonzalez-Gaitan M (2012) Investigating the principles of morphogen gradient formation: from tissues to cells. Curr Opin Genet Dev 22:527–532

8. Pellegrini L, Burke DF, von Delft F, Mulloy B, Blundell TL (2000) Crystal structure of fibroblast growth factor receptor ectodomain bound to ligand and heparin. Nature 407: 1029–1034

9. Chu H, Gao J, Wang Y (2012) Design, synthesis, and biocompatibility of an arginine-based polyester. Biotechnol Prog 28:257–264

10. Daye LR, Gibson W, Williams KP (2010) Development of a high throughput screening assay for inhibitors of hedgehog-heparin interactions. Int J High Throughput Screen 1:69–80

11. Johnson NR, Wang Y (2013) Controlled delivery of sonic hedgehog morphogen and its potential for cardiac repair. PLoS One 8:e63075

12. Chu H, Chen CW, Huard J, Wang Y (2013) The effect of a heparin-based coacervate of fibroblast growth factor-2 on scarring in the infarcted myocardium. Biomaterials 34:1747–1756

13. Chu H, Gao J, Chen CW, Huard J, Wang Y (2011) Injectable fibroblast growth factor-2 coacervate for persistent angiogenesis. Proc Natl Acad Sci U S A 108:13444–13449

14. Chu H, Johnson NR, Mason NS, Wang Y (2011) A [polycation:heparin] complex releases growth factors with enhanced bioactivity. J Control Release 150:157–163

15. Johnson NR, Wang Y (2013) Controlled delivery of heparin-binding EGF-like growth factor yields fast and comprehensive wound healing. J Control Release 166:124–129

16. Li H, Johnson NR, Usas A, Lu A, Poddar M, Wang Y, Huard J (2013) Sustained release of bone morphogenetic protein 2 via coacervate improves the osteogenic potential of muscle-derived stem cells. Stem Cells Transl Med 2:667–677

17. Lee KW, Johnson NR, Gao J, Wang Y (2013) Human progenitor cell recruitment via SDF-1alpha coacervate-laden PGS vascular grafts. Biomaterials 34:9877–9885

18. Awada HK, Johnson NR, Wang Y (2014) Dual delivery of vascular endothelial growth factor and hepatocyte growth factor coacervate displays strong angiogenic effects. Macromol Biosci 14:679–686

19. Johnson NR, Ambe T, Wang Y (2014) Lysine-based polycation:heparin coacervate for controlled protein delivery. Acta Biomater 10: 40–46

Chapter 2

In Vivo Imaging of Hedgehog Transport in *Drosophila* Epithelia

Irene Seijo-Barandiarán, Isabel Guerrero, and Marcus Bischoff

Abstract

The Hedgehog (Hh) signaling pathway is a regulator of patterning, cell migration and axon guidance during development as well as of homeostatic events in adult organs. It is highly conserved from *Drosophila* to humans. In many contexts during development, Hh appears to function as a morphogen; it spreads from producing cells to trigger concentration dependent responses in target cells, leading to their specification. During production, Hh undergoes two lipid modifications resulting in a highly hydrophobic molecule. The processes that create lipid-modified Hh for release from producing cells and that move it to target cells in a graded manner are complex. While most of the work done trying to explain Hh gradient formation is based on immunohistochemical studies in steady state, in vivo imaging in intact organisms is the finest technique to study gradient formation in real time. Both the wing imaginal disc epithelium and the adult abdominal epidermis of *Drosophila* are well suited for in vivo imaging. They allow us to observe the behavior of cells and fluorescently labeled proteins, without interfering with development. Here, we describe in vivo imaging methods for these two epithelia, which allowed us to study Hh transport along specialized cytoplasmic protrusions called cytonemes.

Key words In vivo imaging of Hedgehog signaling, Morphogen dispersion in *Drosophila*, Cell communication, Cytonemes, Exosomes

---

1 Introduction

Hh is essential for the morphogenesis of most organs in the fly and for many metazoan tissues. During development, Hh acts as a morphogen in many contexts. Hh disperses from producing to receiving cells in a graded manner, activating different target genes depending on the concentration they receive. Thus, cells close to the source of production will receive higher Hh levels than cells far from the source, and this will regulate tissue growth and cell fate specification [1]. The mechanism by which morphogens are transported from producing

---

Electronic supplementary material: The online version of this chapter (doi:10.1007/978-1-4939-2772-2_2) contains supplementary material, which is available to authorized users.

Natalia A. Riobo (ed.), *Hedgehog Signaling Protocols*, Methods in Molecular Biology, vol. 1322, DOI 10.1007/978-1-4939-2772-2_2, © Springer Science+Business Media New York 2015

to receiving cells and how morphogen gradients are formed is still under debate [2, 4, 5].

Hh is posttranslationally modified by the addition of cholesterol [6] and palmitic acid [7]. This confers high hydrophobicity to this signaling protein and tethers it to lipid membranes, thus impairing its spreading by free diffusion [6]. There are different models that explain how lipid-modified Hh is secreted and transported from producing to receiving cells: restricted diffusion [8], planar transcytosis [9], argosomes [10], exosomes [11, 12], lipoprotein particles [13], and cellular protrusions, called cytonemes [3, 14, 15].

The mechanisms of Hh gradient formation have been studied extensively in the *Drosophila* wing imaginal disc. The wing disc is a flattened sac made of two layers of closely juxtaposed polarized epithelial cells—the columnar cells of the disc proper and the squamous cells of the peripodial membrane (Fig. 1a, b). Four cell populations with different affinities divide the disc proper in four compartments: anterior (A), posterior (P), dorsal (D), and ventral (V) [16]. Hh is produced in the P compartment and moves across the A/P compartment border decreasing in concentration as it spreads into the A compartment (Fig. 1b) [6, 17], where it patterns the central region of the wing [18, 19]. The Hh receptor Patched (Ptc) is a readout of Hh signaling; it is upregulated in a graded manner in the A compartment, depending on the graded distribution of the Hh signal [20, 21].

In the larval and adult abdominal epidermis, Hh also acts as a morphogen and is expressed in the P compartment [22, 23]. The adult abdominal epidermis is formed during metamorphosis, when the adult histoblasts replace the larval epithelial cells (LECs) [24]. It comprises a sequence of successive segments with Ptc being expressed in A compartment cells anterior and posterior to each P compartment (Fig. 1c, d). The pattern of Ptc expression in the histoblasts is only established during morphogenesis of the adult tissue [3, 22, 23, 25]. Both the larval and the adult abdominal epidermis of *Drosophila* are excellent systems for in vivo imaging [26], with the adult fly hatching after up to 30 h of imaging.

The behavior of cells and proteins during cell-to-cell communication, in wild type or under different mutant conditions, has mainly been studied using in vitro and ex vivo techniques, which are based on fixation and immunostaining. These techniques can only give a static view and also may lead to variable results, due to changeable conditions during fixation and staining. In vivo approaches, on the other hand, allow the observation of processes that happen in the context of a living organism without the need of fixation and staining. Importantly, they also permit the analysis of changes over time (e.g., changes in cell behavior, protein concentration or distribution).

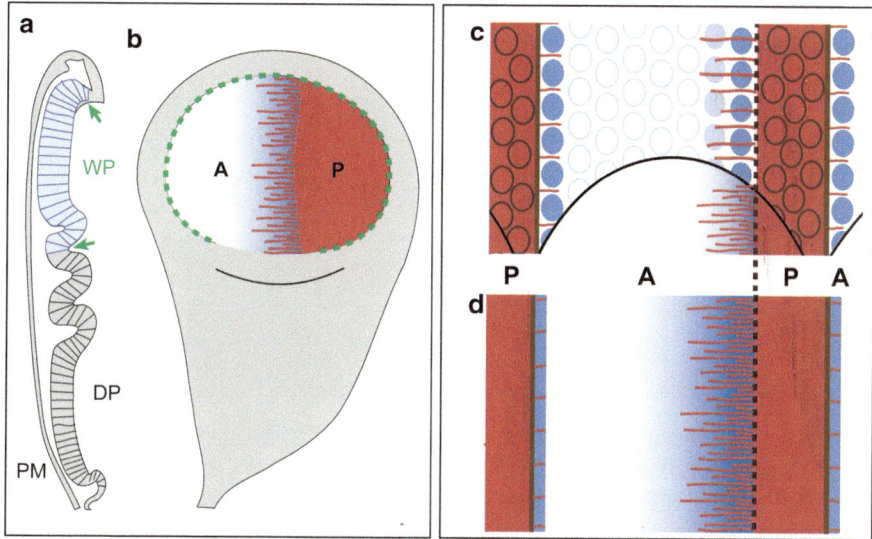

Fig. 1 Schematic representation of Hh signaling in the *Drosophila* wing imaginal disc and abdominal epidermis. (**a, b**) Schemes depict Hh expression domain and the Hh morphogenetic gradient in the wing imaginal disc. (**a**) Transversal section of the wing imaginal disc showing the two juxtaposed epithelia: squamous peripodial membrane (PM) and columnar disc proper (DP). (**b**) Longitudinal DP section of the wing imaginal disc showing the P compartment (*red*) that expresses Hh, which disperses towards the A compartment (*blue*) in a graded manner. This leads to high concentrations of Hh close to the A/P compartment border and lower concentrations far from the source of production. The wing pouch (WP) is delimited by *green arrows* in A and by *green dotted line* in **b**. Note cytonemes emanating from the P compartment (*red*) are present in areas of Hh signaling of the A compartment. (**c, d**) Scheme depicting Hh signaling and cytonemes in one abdominal segment (after Bischoff et al. [3]). Hh and Ptc expression in histoblasts and LECs of A and P compartment (*red*) is shown. (**c**) During histoblast spreading, both histoblasts and LECs express Hh in the P compartment. The Ptc gradient can be observed in the A compartment in both histoblasts and LECs as well as in a row of LECs at the segment boundaries. Cytonemes are present in areas of Hh signaling. (**d**) Adult tissue. Hh is expressed in the P compartment. Ptc is expressed in a gradient in the A compartment as well as in a narrow strip at the segment boundaries. Cytonemes are present in areas of long-range Hh signaling, but not at the segment boundaries, where only local signaling is occurring

In this chapter, we describe in vivo imaging techniques for a better characterization of cell and protein behavior in the context of cell-to-cell communication, namely the imaging of wing imaginal discs, both ex vivo and within the living larva, and the abdominal epidermis (Fig. 2). We have used these techniques to analyze the formation of the Hh morphogen gradient and how exosomes containing Hh move along cytonemes from Hh producing cells to Hh receiving cells [3, 12]. We showed in vivo that cytoneme formation and the establishment of the Hh morphogen gradient correlate in space and time [3]. Furthermore, we showed that Hh-containing exovesicles move along these filopodia-like protrusions [12].

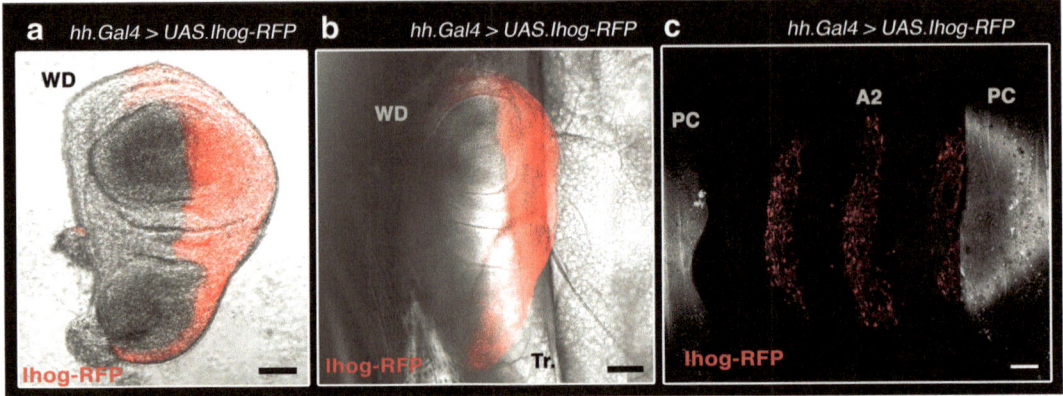

Fig. 2 In vivo imaging of third instar larval wing imaginal disc and pupal abdominal epidermis. In each instance, the P compartments are labeled with *hh.Gal4 > UAS.Ihog-RFP*. Anterior is to the *left*. (**a**) Ex vivo wing disc (WD). Merge of transmitted light and RFP channel. (**b**) In vivo wing disc (WD) in third instar larva. The second thoracic lateral tracheal branch (Tr.) of the larva is also visible. Merge of transmitted light and RFP channel. (**c**) Dorsal view of a pupa, in which a window has been made in the pupal case (PC) for in vivo imaging. The P compartments of segments A1, A2, and A3 are shown

2 Materials

2.1 Ex Vivo Imaging of Drosophila Wing Imaginal Discs

1. *Drosophila* third instar larvae grown using standard protocols [27].
2. 1× PBS buffer: 4.3 mM Na_2HPO_4, 137 mM NaCl, 2.7 mM KCl, 1.4 mM KH_2PO_4.
3. Media composition: M3 insect medium was supplemented with 2 % FBS (Invitrogen #10106-169) and 0.5 % penicillin–streptomycin (Invitrogen #15140-122) and 0.1 µg/ml 20-hydroxyecdysone (Sigma-Aldrich #H5142). Reserve half of the prepared medium to add methylcellulose (M0387-100G; Sigma) at a concentration of 2.5 % wt/vol.
4. Chamber: standard microscope slide, double-sided sticky tape, 24 × 40 mm and 15 × 15 mm coverslips (both No. 1).
5. Poly-lysine (Sigma-Aldrich #P6282) and sterile water.
6. Forceps and needles.
7. A confocal laser-scanning microscope with optics for imaging bright field and fluorescence, such as Zeiss LSM710 upright.

2.2 In Vivo Imaging of Drosophila Wing Imaginal Discs in Living Larvae

1. *Drosophila* third instar larvae from fly stocks grown using standard protocols [27].
2. 1× PBS buffer.
3. Liquid glue: heptane and double-sided sticky tape.
4. Voltalef oil 10 S (VWR International, Cat No. 24627188).
5. Chamber: standard microscope slide, double-sided sticky tape, 24 × 40 and 20 × 20 mm coverslips (both No. 1).

6. Forceps and needles.

7. A confocal laser-scanning microscope with optics for imaging bright field and fluorescence, such as Zeiss LSM710 upright.

2.3 In Vivo Imaging of Drosophila Pupal Abdomen

Preparation of Pupae for Imaging

1. *Drosophila* pupae from fly stocks grown using standard protocols [27]. Staging of pupae is performed after Bainbridge and Bownes [28].

2. Deionized water.

3. Paintbrush with cutoff bristles.

4. Double-sided sticky tape.

5. Standard microscope slide.

6. Forceps and micro knife (or hypodermic needle).

7. 200 µl pipette with tip.

8. Dissecting microscope (e.g., Leica S6E).

Preparation of Imaging Chamber

1. Standard microscope slide and coverslip (22×50 mm, No. 1).

2. Parafilm M.

3. Vaseline (pure petroleum jelly); melted in a heating block (e.g., Techne Dry-Block).

4. Paint brush.

5. Forceps.

6. Deionized water.

7. Whatman paper (3MM).

8. Humid chamber: 90 mm diameter petri dish with wet tissue, sealed with Parafilm M.

9. Voltalef oil 10 S (VWR International, Cat No. 24627188).

10. A confocal laser-scanning microscope, such as Leica SP8.

3 Methods

For in vivo imaging, protein components of exosomes or cytonemes or proteins implicated in the release, transport, or reception of Hh, were fused to a fluorescent tag. These proteins were expressed in a restricted area of the wing imaginal disc or in a group of histoblasts and LECs under the control of a P compartment specific driver, using the GAL4-UAS system [29]. A good tool for visualizing cytonemes is the expression of fluorescently labeled Interference Hedgehog (Ihog), an Hh co-receptor, which stabilizes cytonemes in the basal parts of wing disc and abdominal epithelia [3, 12, 30, 31]. In addition, membrane markers, such as CD4-Tomato and Myr-RFP, are also good markers to visualize cytonemes and exosomes [3].

Isolated wing discs, third instar larvae, and pupae are all imaged in small chambers. These are assembled using microscope slides and/or coverslips as bottoms and lids, with various spacers in between.

3.1 Ex Vivo Imaging of Drosophila Wing Imaginal Discs

1. Making a chamber (on the day before imaging): use one layer of double-sided sticky tape to surround a 5×5 mm area on a 24×40 mm coverslip. Put a drop of poly-lysine on the area (*see* **Note 1**). Once the drop has dried, wash the area with sterile water and keep the chamber at 4 °C.

2. The next steps are done at 4 °C on ice (*see* **Note 2**). Place *Drosophila* third instar larvae in 1× PBS buffer and wash three times with 1× PBS.

3. Transfer larvae to M3 medium and dissect wing imaginal discs. Place the dissected discs in 25 µl M3 medium in the 5×5 mm area of the chamber and add another 25 µl M3 medium with methylcellulose (*see* **Note 3**). Finally, close the chamber with a 15×15 mm coverslip and place it on a standard microscope slide.

4. Image wing discs with a confocal laser-scanning microscope (*see* **Note 4**) (Fig. 3a).

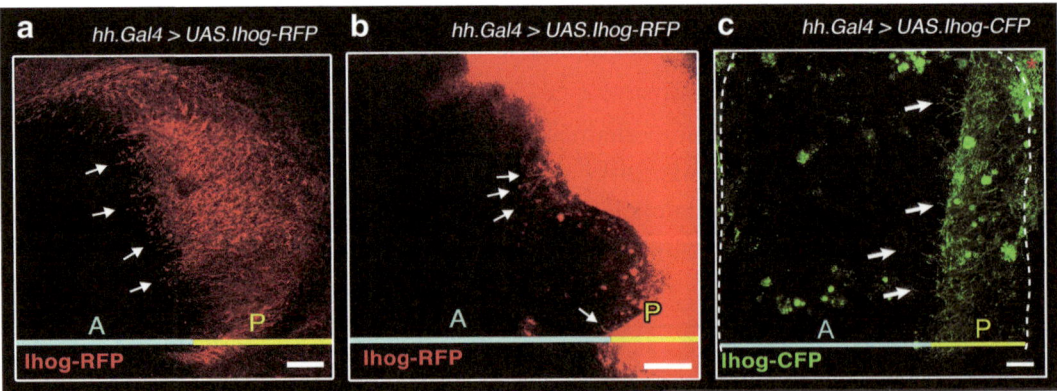

Fig. 3 Cytonemes in wing imaginal discs and the abdominal epidermis. Cytonemes are visualized by driving fluorescently labeled *Ihog* in the P compartment. A and P compartments indicated by *cyan* and *yellow lines*, respectively. Bars, 20 µm. Anterior compartment is oriented to the *left*. (**a**) Basal view of the disc proper of an ex vivo wing imaginal disc. Cytonemes (*arrows*) grow from P compartment cells towards A compartment cells, thereby crossing the compartment border. (**b**) Basal view of the disc proper of an in vivo disc proper in a living third instar larva. Cytonemes are pointing anteriorly into the A compartment (*arrows*). (**c**) Abdominal epithelium approx. 40 h after puparium formation. Histoblasts are labeled with Ihog-CFP. Along the compartment border, cytonemes point anteriorly (*arrows*). Segment A2 shown (indicated by *hatched line*). *Asterisk* indicates a LEC, which has not yet been replaced during morphogenesis

3.2 In Vivo Imaging of Drosophila Wing Imaginal Discs in Living Larvae

1. Making the chamber: use two layers of double-sided sticky tape to surround a 15×15 mm area on a 24×40 mm coverslip.

2. The next steps are done at 4 °C on ice (*see* **Note 2**). Place *Drosophila* third instar larvae in 1× PBS buffer and wash three times with 1× PBS.

3. Put a drop of liquid glue in the 15×15 mm area of the chamber and press the larva on it until the glue gets dry (*see* **Note 5**). Add a drop of Voltalef oil on the larva to avoid desiccation during imaging. Finally, seal the chamber with a 20×20 mm coverslip.

4. Image wing discs with a confocal laser-scanning microscope (*see* **Notes 6** and **7**) (Fig. 3b).

3.3 In Vivo Imaging of Drosophila Pupal Abdomen

Preparation of Pupae for Imaging

1. Collect pupae with wet paintbrush with cutoff bristles from the walls of a culture tube and stick them on double-sided tape on a standard microscope slide with the dorsal side up.

2. Dry pupae for around 30 min at room temperature.

3. Make a small hole in the pupal case in the space between thorax and abdomen dorso-laterally using a micro knife or a hypodermic needle. Start at this hole to peel a window in the pupal case in your area of interest using forceps.

4. Add some water on the pupae with the pipette and remove them carefully from the sticky tape with forceps to transfer them to the imaging chamber.

Preparation of Imaging Chamber and Imaging

1. Stick two strips of Parafilm "M" spacers (50×4 mm; six layers thick) on the long sides of a standard microscope slide (*see* **Note 8**).

2. Put pupae in the center of the slide.

3. Add two small pieces (ca. 5×5 mm) of wet Whatman paper in the vicinity of the pupae to provide humidity.

4. Add some water to cover the pupae.

5. Put a coverslip (22×50 mm, No. 1) on the spacers; fix it with melted Vaseline all way along the spacers using a paintbrush.

6. Seal the two open sides of the chamber with Voltalef oil.

7. Image pupae with confocal laser scanning microscope (*see* **Note 9**) (Fig. 3c; Supplementary Video 1).

8. Remove Voltalef oil after filming and keep slide in a humid chamber until flies hatch.

4 Notes

1. We use the poly-lysine to prevent imaginal disc movement during imaging, as it will attach the wing imaginal discs to the coverslip.

2. Working at 4 °C will confer a better manipulation of larvae, as they will not move. This condition also prevents medium contamination.

3. Methylcellulose is used to make the medium more viscous.

4. The wing discs survive imaging for up to 14 h under the used conditions. We use Caspase3 immunostaining after imaging to check that cells do not undergo apoptosis due to the culture conditions.

5. The liquid glue is made of double-sided sticky tape that is kept overnight in a bottle of heptane until all the glue has dissolved in the heptane. Then the tape is removed. The glue is used to prevent larval movement in the imaging chamber as much as possible. Depending on the manufacturer of the sticky tape, the glue might be toxic to the larvae. Therefore, it might be necessary to trial a few different tapes.

6. Wing discs expressing a fluorescent protein are easily visualized in the living larvae situated on the second thoracic lateral tracheal branch. Due to interference of the larval integument and tissue, imaging of cellular structures in the wing imaginal disc is challenging and high laser power is required. Imaginal discs that are positioned close to the epidermis with little tissue below are best to image.

7. Due to residual larval movement, time-lapse imaging is limited to a few minutes under these conditions. Larvae can be kept in the imaging chambers for around 30 min. After imaging, the larvae are transferred into standard fly food vials, in which they often pupate and produce adult flies, indicating that imaging is not impairing larval development.

8. The number of layers of Parafilm M can be adjusted depending on the size of the pupae, which might vary depending on growing conditions and genotype.

9. Due to the nature of the chamber, there might be considerable drift of the plane of focus during the course of a time-lapse recording. We take this into account by starting the recording 30 μm above the top of the specimen.

Acknowledgements

We are very grateful to Carmen Ibáñez for technical assistance. We also thank the confocal microscopy facility of the CBMSO for skillful technical assistance. Work was supported by grant BFU2011-

25987 from the Spanish MINECO and by an institutional grant to the CBMSO from the Fundación Areces to I.G. I.S. was financially supported by an FPI fellowship of the Spanish MINECO.

References

1. Briscoe J, Therond PP (2013) The mechanisms of Hedgehog signalling and its roles in development and disease. Nat Rev Mol Cell Biol 14:416–429

2. Guerrero I, Kornberg TB (2014) Hedgehog and its circuitous journey from producing to target cells. Semin Cell Dev Biol 33C:52–62

3. Gradilla AC, Guerrero I (2013) Hedgehog on the move: a precise spatial control of Hedgehog dispersion shapes the gradient. Curr Opin Genet Dev 23:363–373

4. Ingham PW, Nakano Y, Seger C (2011) Mechanisms and functions of Hedgehog signalling across the metazoa. Nat Rev Genet 12:393–406

5. Porter JA, Young KE, Beachy PA (1996) Cholesterol modification of hedgehog signaling proteins in animal development. Science 274:255–259

6. Pepinsky RB, Zeng C, Wen D, Rayhorn P, Baker DP, Williams KP, Bixler SA, Ambrose CM, Garber EA, Mlatkowski K, Taylor FR, Wang EA, And Galdes A (1998) Identification of a palmitic acid-modified form of human Sonic hedgehog. J Biol Chem 273:14037–14045

7. Crick F (1970) Diffusion in embryogenesis. Nature 225:420–422

8. Entchev EV, Schwabedissen A, Gonzalez-Gaitan M (2000) Gradient formation of the TGF-beta homolog Dpp. Cell 103:981–991

9. Greco V, Hannus M, Eaton S (2001) Argosomes: a potential vehicle for the spread of morphogens through epithelia. Cell 106:633–645

10. Gross JC, Chaudhary V, Bartscherer K, Boutros M (2012) Active Wnt proteins are secreted on exosomes. Nat Cell Biol 14:1036–1045

11. Gradilla A-C, González E, Seijo I, Andrés G, Bischoff M, González-Mendez L, Sánchez V, Callejo A, Ibáñez C, Guerra M, Ortigão-Farias J-R, Sutherland J-D, González M, Barrio R, Falcón-Pérez J-M, Guerrero I (2014) Exosomes as Hedgehog carriers in cytoneme-mediated transport and secretion. Nat Commun 5:5649. doi:10.1038/ncomms6649

12. Panakova D, Sprong H, Marois E, Thiele C, Eaton S (2005) Lipoprotein particles are required for Hedgehog and Wingless signalling. Nature 435:58–65

13. Bischoff M, Gradilla AC, Seijo I, Andrés G, Rodríguez-Navas C, González-Méndez L, Guerrero I (2013) Cytonemes are required for the establishment of a normal Hedgehog morphogen gradient in *Drosophila* epithelia. Nat Cell Biol 15:1269–1281

14. Ramirez-Weber FA, Kornberg TB (1999) Cytonemes: cellular processes that project to the principal signaling center in *Drosophila* imaginal discs. Cell 97:599–607

15. Roy S, Hsiung F, Kornberg TB (2011) Specificity of *Drosophila* cytonemes for distinct signaling pathways. Science 332:354–358

16. Garcia-Bellido A, Ripoll P, Morata G (1973) Developmental compartmentalisation of the wing disk of *Drosophila*. Nat New Biol 245: 251–253

17. Tabata T, Kornberg TB (1994) Hedgehog is a signaling protein with a key role in patterning *Drosophila* imaginal discs. Cell 76:89–102

18. Mullor JL, Calleja M, Capdevila J, Guerrero I (1997) Hedgehog activity, independent of decapentaplegic, participates in wing disc patterning. Development 124:1227–1237

19. Strigini M, Cohen SM (1997) A Hedgehog activity gradient contributes to AP axial patterning of the *Drosophila* wing. Development 124:4697–4705

20. Capdevila J, Estrada MP, Sanchez-Herrero E, Guerrero I (1994) The *Drosophila* segment polarity gene patched interacts with decapentaplegic in wing development. EMBO J 13:71–82

21. Chen Y, Struhl G (1996) Dual roles for patched in sequestering and transducing Hedgehog. Cell 87:553–563

22. Struhl G, Barbash DA, Lawrence PA (1997) Hedgehog organises the pattern and polarity of epidermal cells in the *Drosophila* abdomen. Development 124:2143–2154

23. Kopp A, Muskavitch MA, Duncan I (1997) The roles of hedgehog and engrailed in patterning adult abdominal segments of *Drosophila*. Development 124:3703–3714

24. Madhavan MM, Madhavan K (1980) Morphogenesis of the epidermis of adult abdomen of *Drosophila*. J Embryol Exp Morphol 60:1–31

25. Kopp A, Duncan I (2002) Anteroposterior patterning in adult abdominal segments of *Drosophila*. Dev Biol 242:15–30

26. Bischoff M, Cseresnyes Z (2009) Cell rearrangements, cell divisions and cell death in a migrating epithelial sheet in the abdomen of *Drosophila*. Development 136:2403–2411

27. Ashburner M, Roote J (2007). Maintenance of a *Drosophila* laboratory: general procedures. CSH Protoc 2007: pdb.ip35

28. Bainbridge SP, Bownes M (1981) Staging the metamorphosis of *Drosophila* melanogaster. J Embryol Exp Morphol 66:57–80

29. Brand AH, Perrimon N (1993) Targeted gene expression as a means of altering cell fates and generating dominant phenotypes. Development 118:401–415

30. Bilioni A, Sánchez-Hernández D, Callejo A, Gradilla AC, Ibáñez C, Mollica E, Carmen Rodríguez-Navas M, Simon E, Guerrero I (2013) Balancing Hedgehog, a retention and release equilibrium given by Dally, Ihog, Boi and shifted/DmWif. Dev Biol 376:198–212

31. Callejo A, Billioni A, Mollica E, Gorfinkiel N, Andrés G, Ibáñez C, Torroja C, Doglio L, Sierra J, Guerrero I (2011) Dispatched mediates Hedgehog basolateral release to form the long-range morphogenetic gradient in the *Drosophila* wing disk epithelium. Proc Natl Acad Sci U S A 108:12591–12598

Chapter 3

Modeling Hedgehog Signaling Through Flux-Saturated Mechanisms

Óscar Sánchez, Juan Calvo, Carmen Ibáñez, Isabel Guerrero, and Juan Soler

Abstract

Hedgehog (Hh) molecules act as morphogens directing cell fate during development by activating various target genes in a concentration dependent manner. Hitherto, modeling morphogen gradient formation in multicellular systems has employed linear diffusion, which is very far from physical reality and needs to be replaced by a deeper understanding of nonlinearities. We have developed a novel mathematical approach by applying flux-limited spreading (FLS) to Hh morphogenetic actions. In the new model, the characteristic velocity of propagation of each morphogen is a new key biological parameter. Unlike in linear diffusion models, FLS modeling predicts concentration fronts and correct patterns and cellular responses over time. In addition, FLS considers not only extracellular binding partners influence, but also channels or bridges of information transfer, such as specialized filopodia or cytonemes as a mechanism of Hh transport. We detect and measure morphogen particle velocity in cytonemes in the Drosophila wing imaginal disc. Indeed, this novel approach to morphogen gradient formation can contribute to future research in the field.

Key words Hedgehog gradient, Mathematical modeling, Cytonemes and vesicles in Hh, Flux-saturated mechanism, Dispersion

1 Introduction

An important question in biology is to understand how secreted morphogens provoke distinct cell fates in a concentration-dependent way. Hedgehog (Hh) is one of the most important morphogens to regulate growth, cell fate specification and cell migration during development, and also to maintain tissue homeostasis (reviewed in ref. [1]). The mechanism of signal distribution from the producing to the receiving cells is less known about morphogens gradient formation. Several mathematical models have been proposed to explain morphogen movement.

Mathematical models need to recapitulate, to detail enabled by technology, events experimentally observed. Moreover, the model ability to predict nontrivial conditions confers its biological relevance.

Natalia A. Riobo (ed.), *Hedgehog Signaling Protocols*, Methods in Molecular Biology, vol. 1322,
DOI 10.1007/978-1-4939-2772-2_3, © Springer Science+Business Media New York 2015

Indeed, prediction usually goes beyond than just accurately mimicking known experimental results. Mathematics apply to morphogenesis have been classically based on reaction–diffusion models since the pioneering work of Turing, Crick and Meinhardt [2–6]. Modeling morphogenetic processes has taken a well-known path, largely playing with parameters within Reaction-Diffusion systems, in order to get a quantitative fitting to reproduce previously established patterns. Quantitative accuracy is important, but qualitative aspects of morphogen gradient evolution are no less important; they might shed light on morphogenetic processes and validate a mathematical model as a driving force for new biological predictions. On large time scales, qualitative and quantitative properties must converge, but often what happens at small scales affects critically to pattern formation, as we shall see in the case of the Sonic Hedgehog (Shh) signaling pathway and its impact on the activation of the target gene *gli*.

We have focused here in the dynamic properties of signaling pathway Shh-Gli, which is involved in embryonic development in vertebrate, when Gli is regulated, as well as in tumor progression when Gli is deregulated [7, 8]. In the developmental biology context, the vertebrate neural tube, the chick limb bud, and the insect wing imaginal disc (where the morphogen Hh plays a similar role than Shh in vertebrates) are prime examples of morphogenetic patterning. In the former, embryonic pseudostratified ventral epithelial cells are instructed to acquire specific neural fates in response to Shh morphogenetic signal derived from the ventral floor plate, and earlier from the notochord [9]. In the latter, Hh signal is produced in the posterior compartment to establishes patterns in the anterior compartment of the developing pseudostratified wing epithelium. In all model cells closest to the source receive higher doses of a morphogen and for longer times than those further apart from it, but, importantly, distal-most cells do not receive Hh.

In the above examples reactions applying linear diffusion systems do not collect the real biological features. In fact, a number of biological properties are incompatible with linear diffusion. First, Hh morphogen aggregates are not small particles in large spaces, thus invalidating one of the assumptions of Brownian motion giving rise to linear diffusion: Hh has been detected in *Drosophila* as aggregates of 20–300 nm in diameter [10, 11]. Such particles have been shown to be in oligomers [11], multimers [12], membrane vesicles [13], and/or lipoprotein particles [14, 15]. Second, morphogen dynamics to induce pattern formation are conflicting with the action of linear diffusion because the time of exposure is critical to specify cell fate.

Modeling morphogen gradients by using linear diffusion necessarily implies that in the responding field every cell receives a certain level of morphogen instantaneously and, therefore, foresees the same time of exposure for every cell in the whole tissue [7]. This is at odds as development of morphogenetic responses requires time. Indeed, low level Shh signaling has cumulative effects in the neural tube and limb buds during development. Similarly, while current linear diffusion models may reproduce normal final patterns they cannot account for how these are formed [16]. Indeed, in order to make the concentration curves fit with the biological reality of a discontinuity at the front of the gradient an arbitrary threshold of non-instructive signaling or noise had to be introduced in such models (*see* Fig. 8 in ref. [17]). Cutting the tails of the Gaussian-type solution is a usual procedure in Reaction-Diffusion systems, in particular in traveling waves solutions, by arguing a good accuracy with zero and by trying to overcome the nonrealistic infinite character of the tails. However, this adjustment produces artificial fronts without connection with the biological context. It is important to note that the existence of a front and also the velocity of morphogen transport (which obviously is not infinite, contrary to what corresponds to linear diffusion), are both critical elements in morphogenetic processes. Without a threshold mechanism a chemical signal arrives to distant areas too fast, thus triggering the chemical cascade too soon. In this way, previous works based on linear diffusion simply apply known data, but have to impose non-biological thresholds in order to have some sense, namely for traveling fronts [17]. These models, then, could not predict the novelty introduced by biological processes, such as the effect of a negative feedback [7].

Moreover, such linear diffusion models do not appropriately account for the anatomy where the macroparticles (vesicles) are being transported. Thus, they do not account for the topography of its signaling landscape, i.e., the precise physical relationships between the cells that make signaling proteins and the fields of cells that receive these signals [18]. The proposal of specialized filopodia, also called cytonemes, [19] for morphogens transport satisfies both and is the most closest to present experiment mechanism of protein transport. Cytonemes are actin-based cell protrusions and span several cell diameters. Various Hh signaling components localize to cytonemes, as well as to vesicle-like structures (macroparticles) moving along cytonemes [20–22] (*see* Fig. 1). Cytonemes are dynamic structures and Hh gradient is established and correlated with cytoneme formation in space and time. This cytoneme-based model confronts the previous diffusion-based models [18, 19, 23–25]. In general, models that cannot accommodate all of the different settings have limited biological relevance.

In addition, our model using flux-saturated mechanism [26] indeed predicts nontrivial behavior as it incorporates such interaction mechanisms between fluxes and target genes. In the Hh signaling, our model also takes into account the action of the Hh receptor Ptc1, introducing a negative feedback component. Therefore, our proposed model is an extension to Fick's law and of the pioneering ideas of Turing, but overcomes the qualitative

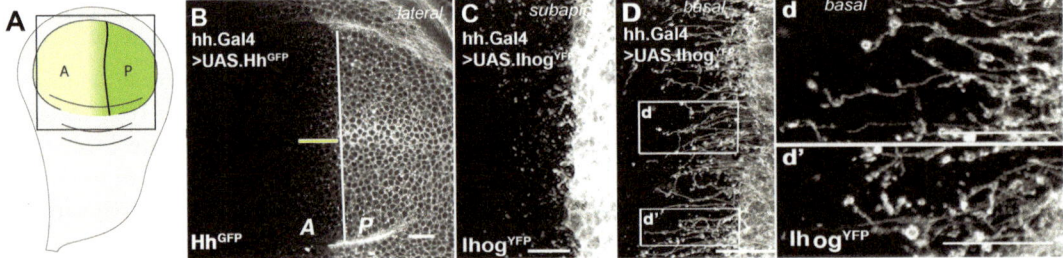

Fig. 1 Hh and its co-receptor Ihog are present in punctate structures that also localize to cytonemes. (**a**) Schematic representation of a wing imaginal disc. The *central colored region* corresponds to the wing primordium (wing pouch), where the expression of Hh in the P compartment, and the Hh gradient at the A/P compartment border are in *green*. (**b**) A lateral confocal section of a hh.Gal4 > UAS.HhGFP wing disc showing the localization of Hh in the P compartment of the wing pouch and *dots* in the A compartment that correspond to Hh gradient. (**c**) A subapical section of an hh.Gal4 > UAS.Ihog^YFP wing disc. Very short cytonemes and also punctate structures in the A compartment cells can be observed. (**d**) Most basal confocal section of a hh.Gal4 > UAS.Ihog^YFP wing disc. Note in the inset (*d, d'*) that punctate structures labeled by Ihog^YFP are attached to cytonemes. *Bars*, 10 μm

weaknesses associated with linear diffusion. Our current analysis to develop a general model for Shh-Gli signaling in the vertebrate neural tube is based on a continuous feedback between mathematical modeling, numerical analysis and a collection from experimental data on Hh particle movement in the Drosophila wing imaginal disc epithelium [26]. Our thesis implies a transport of active macroparticles (vesicles) throughout specialized filopodia (cytonemes), where cooperation and strategy are linked, and they are the biological counterparts of the flux-saturated mechanism [26].

As we have pointed out previously, in the Hh pathway the number of aggregates is infinitely large from a scale point of view, trafficking is large, and a molecular description is difficult to model since several molecular and cellular interactions are involved. There are also positive or negatively regulators of the Hh pathway. For instance, the cell adhesion molecule Cdon, and its Drosophila ortholog Ihog, forms an heteromeric complex with the Hh receptor Patched 1 [27, 28]. This receptor complex binds Hh and enhances signaling activation, indicating that Cdon, and Ihog, positively regulate the pathway. There are also evidences indicating that in the absence of Hh–Ptc interaction, Cdon, and also Ihog, have an evolutionary preserved function as Hh decoy receptors [20, 29, 30].

As we also mentioned before the Hh molecules do not travel alone, but they do so in complex aggregates with other molecules, forming vesicles [24, 25], where there are, as we have seen, emerging relations of cooperation and interactions towards a common goal beyond a single behavior. This mechanism of limiting Hh activity acts in parallel to other, more intensively studied mechanisms such as the negative feedback regulation of the Ptc1 receptor, which inhibits pathway activation.

2   Materials

2.1   Fly Stocks

To induce ectopic expression of fluorescent proteins we use the *Gal4/UAS* system [31]. The *pUAS-* transgenes were: UAS-Hh-GFP [32], and UAS-Ihog-YFP [21]. The Gal4 driver stock was *hh-Gal4* [33].

2.2   Ex Vivo Imaging of Drosophila Wing Imaginal Discs

1. *Drosophila melanogaster* third instar larvae grown using standard protocols [34].

2. 1× PBS buffer.

3. Media composition: M3 insect medium was supplemented with 2 % FBS (Invitrogen #10106-169) and 0.5 % penicillin-streptomycin (Invitrogen #15140-122) and 0.1 µg/mL 20-hydroxyecdysone (Sigma-Aldrich #H5142). Reserve half of media to add methyl-cellulose (M0387-100G; Sigma) at a concentration of 2.5 % wt/vol.

4. Chamber: standard microscope slide, double-sided sticky tape, 24×40 mm and 15×15 mm coverslips (both No. 1).

5. Poly-lysine (Sigma-Aldrich #P6282) and sterile water.

6. Forceps and needles.

7. A confocal laser scanning microscope with optics for imaging bright field and fluorescence, such as Zeiss LSM710 upright.

3   Methods and Results

3.1   In Vivo Imaging of Drosophila Wing Imaginal Discs

For in vivo imaging, Hh protein or the Hh co-receptor Ihog, which are located in exosomes or cytonemes were fused to a fluorescent tag [35, 36]. Transient expression of the UAS-constructs, which respond to Gal4 induction, was done by using the Gal4; tub-Gal80^{ts} system and maintaining the fly crosses at 18 °C and inactivating the Gal80^{ts} for 24–30 h at the restrictive temperature (29 °C) before discs dissection. These proteins were expressed in a restricted area of the wing imaginal disc under the control of A/P compartment specific driver (HhGal4).

1. Making a chamber (on the day before imaging): use one layer of double-sided sticky tape to surround a 5×5 mm area on a 24×40 mm coverslip. Put a drop of poly-lysine on the area (*see* **Note 1**). Once the drop has dried, wash the area with sterile water and keep the chamber at 4 °C.

2. The next steps are done at 4 °C on ice (*see* **Note 2**). Place *Drosophila* third instar larvae in 1× PBS buffer and wash three times with 1 × PBS.

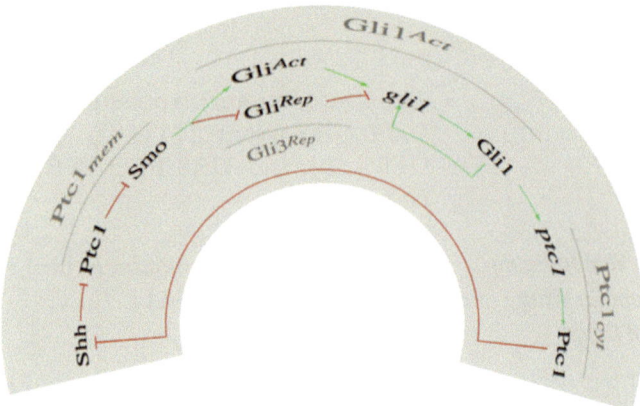

Fig. 2 Fluorescence recovery after photobleaching (FRAP). (**a–f**) Confocal sections of a third instar Drosophila wing imaginal disc expressing Hh-GFP, at time, $t = 0$ (**a**), just after photobleaching, $t = 15$ s (**b**), where the photobleached area is in a *red rectangle*; and at different times after photobleaching, $t = 85, 270, 425, 610$ s (**c–f**). *The twisted red line* indicates the front position of the same confocal section at different time recovery. (**g**) Plot of the values of the average pixel intensity at various times after photobleaching as a function of the distance from the A/P compartment border of the imaginal wing disc. The *arrows* in (**g**) indicate the apparent position of the signaling front. Normalized fluorescence intensities are also shown and determined on very narrow strips with widths ranging between 0.68 and 1.68 μm perpendicular to the front propagation. Data are derived from eight different FRAP experiments and calculated as a function of the distance from the A/P compartment border. The finite velocity of propagation of the Hh fluorescent front, ranges from 0.07 to 0.02 μm/s with a maximum variation of ±0.015 μm/s in each experimental FRAP

3. Transfer larvae to M3 medium and dissect wing imaginal discs. Place the dissected discs in the pre-made chamber (*see* **Note 3**). Finally, close the chamber with a 15×15 mm coverslip and place it on a standard microscope slide.

4. Image wing discs with a confocal laser scanning microscope (as in Fig. 2).

3.2 Time Lapse Confocal Image and FRAP Analyses in the Wing Imaginal Discs

Laser scaning confocal microscopes (Zeiss laser confocal and multiphoton LSM710 coupled to an inverted Axio Observer (Zeiss) and Argon 458/488/514 nm laser and DPSS 561 nm laser) are used for confocal fluorescence imaging. Image J software (NIH USA) was used for image processing and determining fluorescence levels. All image time series are acquired with a 40×/1.30 EC "Plan-Neofluar" Oil DIC M27 (Zeiss) and using a Zeiss Zen 2008 program.

Fluorescence recovery after photobleaching (FRAP) experiments in the third instar *Drosophila* wing imaginal disc using a Hh-GFP transgenic line, at time $t = 0$ (A) and $t = 275$ s is shown in Fig. 2.

3.3 Numerical Solutions of the Flux Limited Spreading (FLS) Model

The numerical solution of the whole system of eight coupled equations, coded in C, considers a space discretization of the flux limiter transport equation for Shh by using a fifth-order finite difference WENO (Weighted Essentially Non-oscillatory) scheme [37] with Lax–Friedrichs flux splitting [38]. For the time evolution a fourth order Runge–Kutta method has been employed [32]. A space grid of 1,000 points and time grid of 7×10^6 points was considered.

3.4 Flux Saturated Mechanisms for the Transport of Shh. Modeling Morphogenetic Responses

In the vertebrate neural tube model, the main objective is trying to understand how morphogen gradients are formed and interpreted (signal transduction by the receiving cells). This model studies the Dorso-Ventral (DV) patterning in the chick embryo spinal cord, beginning when Shh is first secreted by the floor plate. It focuses not on the whole neural tube, but only on the ventral-most binary cell fate (V3 interneurons). The most relevant proteins involved in the transduction process are Ptc and Gli proteins and follow the interaction scheme of Fig. 3; $PtcShh_{mem}$, $PtcShh_{cyt}$, Ptc_{mem}, Ptc_{cyt}, $Gli1^{Act}$, $Gli3^{Act}$, and $Gli3^{Rep}$. In *Drosophila* the interaction scheme is similar, a little more simplified, where the target of the pathway are *ptc* and *ci* (*cubitus interruptus*) instead of *ptc1* and *gli1*.

The problem of infinite speed of propagation for linear diffusion equations dates back to Fick's theory, which is based on a linear relation between the concentration flux and the gradient $\nabla u(t, x)$ of the concentration function. The subsequent macroscopic equation associated to the Fick law gives $\partial_t u = k \operatorname{div}(\nabla u)$, and predicts an infinite speed of propagation for the concentration flux. Flux-limitation mechanisms propose to modify the microscopic law defining the flow to make it saturate when concentration gradients become unbounded. The result is a nonlinear spreading equation of the type

$$
\frac{\partial u}{\partial t} = \nu \operatorname{div}\left(\frac{|u|^m \nabla_x u}{\sqrt{u^2 + \dfrac{\nu^2}{c^2} |\nabla_x u|^2}} \right) + F(u),
$$

where $F(u)$ takes into account the reaction terms involved inside the cells. In the case $m = 1$, this equation with $F(u) = 0$ was first introduced by Rosenau in ref. [39] and later derived by means of optimal mass transportation in ref. [40]. It can also be recovered performing macroscopic limits of kinetic models of multicellular interactions [36]. Here the constant c is the maximum speed of propagation allowed in the medium, while ν stands for a viscosity and reduces to a diffusion coefficient in the limit of the velocity $c \to \infty$, in which the usual Fick equation is recovered.

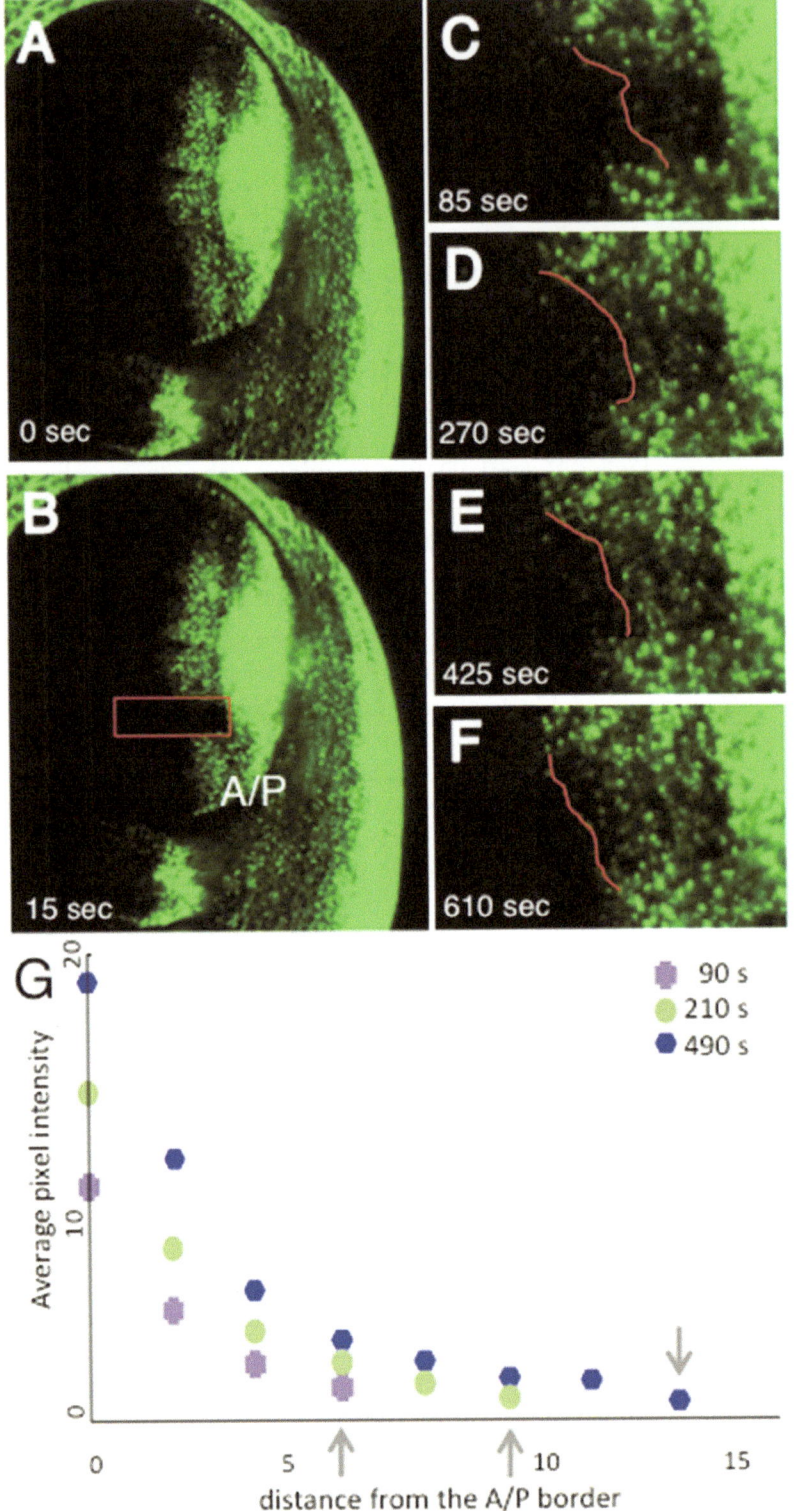

Fig. 3 Schematic diagram of the Shh-Gli pathway. The temporal time line follows the flow of positive *arrows* and negative T bars from the action of the secreted Shh morphogen on the left. Note that the Gli code includes the function of Gli2 and Gli3 activators as well as Gli2 repressors. For simplicity we incorporate all repressor function in the term Gli3[Rep]. For activators, we include Gli3[Act] but consider Gli1[Act] as the major Gli activator

Using this kind of ideas, the following flux-limited spreading (FLS) equation results once we apply the mass action law:

$$\frac{\partial [\text{Shh}]}{\partial t} = v\, \partial_x \left(\frac{[\text{Shh}]^m\, \partial_x [\text{Shh}]}{\sqrt{[\text{Shh}]^2 + \dfrac{v^2}{c^2}\left(\partial_x [\text{Shh}]\right)^2}} \right) + k_{\text{off}}\left[\text{PtclShh}_{\text{mem}}\right] - k_{\text{on}}[\text{Shh}]\left[\text{Ptcl}_{\text{mem}}\right],$$

where square brackets denote concentrations and we will assume that $m = 1$, but we will see how the velocity of discontinuity fronts change and fits the experimental results in *Drosophila* when m is not 1.

The chemical cascade of reactions taking place inside the cells will be described by a different set of ordinary differential equations than those of linear diffusion models [17]. Not only because the chemical signal does not arrive instantaneously to the surface receptors, altering the internal dynamics in a significant way, but also because the synthesis and transport to cell membrane of Ptc1 molecules can take some time. This feature seems to have been overlooked in the previous models and it entails an extra delay for the system of differential equations (which is represented by the parameter τ below). This set of differential equations describing biochemical reactions inside the cells reads now as follows:

$$\frac{\partial\left[\text{PtclShh}_{\text{mem}}\right]}{\partial t} = -\left(k_{\text{off}} + k_{\text{Cin}}\right)\left[\text{PtclShh}_{\text{mem}}\right] + k_{\text{on}}[\text{Shh}]\left[\text{Ptcl}_{\text{mem}}\right] + k_{\text{Cout}}\left[\text{PtclShh}_{\text{cyt}}\right],$$

$$\frac{\partial\left[\text{PtclShh}_{\text{cyt}}\right]}{\partial t} = k_{\text{Cin}}\left[\text{PtclShh}_{\text{mem}}\right] - k_{\text{Cout}}\left[\text{PtclShh}_{\text{cyt}}\right] - k_{\text{Cdeg}}\left[\text{PtclShh}_{\text{cyt}}\right],$$

$$\frac{\partial\left[\text{Ptcl}_{\text{mem}}\right]}{\partial t} = k_{\text{off}}\left[\text{PtclShh}_{\text{mem}}\right] - k_{\text{on}}[\text{Shh}]\left[Ptcl_{\text{mem}}\right] - k_{\text{Pint}}\left[\text{Ptcl}_{\text{cyt}}\right],$$

$$\frac{\partial\left[\text{Ptcl}_{\text{cyt}}\right]}{\partial t} = k_{\text{p}} P_{\text{tr}}\left\{\left[\text{Gli1}^{\text{Act}}\right](t-\tau), \left[\text{Gli3}^{\text{Act}}\right](t), \left[\text{Gli3}^{\text{Rep}}\right](t)\right\}\Phi_{\text{Ptc}} - k_{\text{Pint}}\left[\text{Ptcl}_{\text{cyt}}\right],$$

$$\frac{\partial\left[\text{Gli3}^{\text{Act}}\right]}{\partial t} = k_{\text{G}} P_{\text{tr}}\left\{\left[\text{Gli1}^{\text{Act}}\right](t-\tau), \left[\text{Gli3}^{\text{Act}}\right](t), \left[\text{Gli3}^{\text{Rep}}\right]\right\}\Phi_{\text{Ptc}} - k_{\text{deg}}\left[Gli1^{\text{Act}}\right],$$

$$\frac{\partial\left[\text{Gli3}^{\text{Rep}}\right]}{\partial t} = \left[\text{Gli3}^{\text{Act}}\right]\frac{k_{\text{g3r}}}{1+R_{\text{Ptc}}} - k_{\text{deg}}\left[\text{Gli3}^{\text{Rep}}\right],$$

$$\frac{\partial\left[\text{Gli3}^{\text{Act}}\right]}{\partial t} = \frac{\gamma_{\text{g3}}}{1+R_{\text{Ptc}}} - \left[Gli^{\text{Act}}\right]\frac{k_{\text{g3r}}}{1+R_{\text{Ptc}}} - K_{\text{deg}}\left[\text{Gli3}^{\text{Act}}\right],$$

being

$$\Phi_{Ptc} = \frac{\left[Ptcll_0\right]}{\left[Ptcll_0\right] + Ptcll_{mem}}, \quad R_{Ptc} = \frac{\left[PtclShh_{mem}\right]}{\left[Ptcl_{mem}\right]},$$

where $[Ptcl_0]$ is the initial value of $[Ptcl_{mem}]$ and the weight term P_{tr} evaluates the probability corresponding to transcriptional events occurring through the interactions of regulatory proteins ($Gli1^{Act}$, $Gli3^{Act}$ and $Gli3^{Rep}$) with specific DNA sequences (the target genes *gli1* and *ptc1*) (*see* **Note 4**). The values for the many parameters above are either taken from the literature or obtained experimentally [25]. From now on, we will refer to the coupling of the FLS equation with the ODEs system as the Gli-FLS model. The mixed Dirichlet-Neumann problem (the well-posedness as well as the asymptotic behavior of the solutions) for the FLS equation has been analyzed in refs. [26, 37, 38, 41]. Interestingly enough, the velocity at which the incoming chemical signal (vesicles) travels through the cytonemes is finite, which is precisely the behavior that we wanted to describe with a mathematical model, and which cannot be attained using a model like that in ref. [17]. The value of c can be measured experimentally in different systems [26]. The mean value in the experiments for c was found to be 0.07 μm/s in the first few seconds followed by an average value of 0.02 μm/s. Other measures in wing discs (8 h after Hh-GFP induction) and in early (embryonic day 8.5) mouse neural tubes, however, suggest a value of 0.0013 μm/s. The value of c thus appears to be variable in different contexts (*Drosophila*, culture cells or vertebrates) and at different times or stages of development and in areas of equivalent distance from the source, *see* ref. [26]. The speed variation at different stages of morphogen transport allows us to elucidate that a model in which the flux saturation combined with the porous nature ($m > 1$) of the extracellular medium could improve the quantitative and qualitative aspects of the model (Fig. 4).

3.5 Future Perspectives

As we have pointed out before, the mechanism of morphogen movement from producing to receiving cells has been a subject of intense study from both mathematical and biological points of view. The transport of the information by specialized filopodia or cytonemes that orient towards the morphogenetic source and vice versa, to the receiving cells, is at the present the mechanism with more biological evidences [18, 19, 24, 30, 35]. Therefore, the way in which these macromolecules are transported is not random. Furthermore, the formation, evolution and transfer of information by cytonemes and vesicles that move along cytonemes is thus a complex process which would involve ligand–receptor interaction, which in turn also may determine the directionality and dynamic of cell extensions. It has been also proposed that the exchange of information could be carried across specialized cellular synapses,

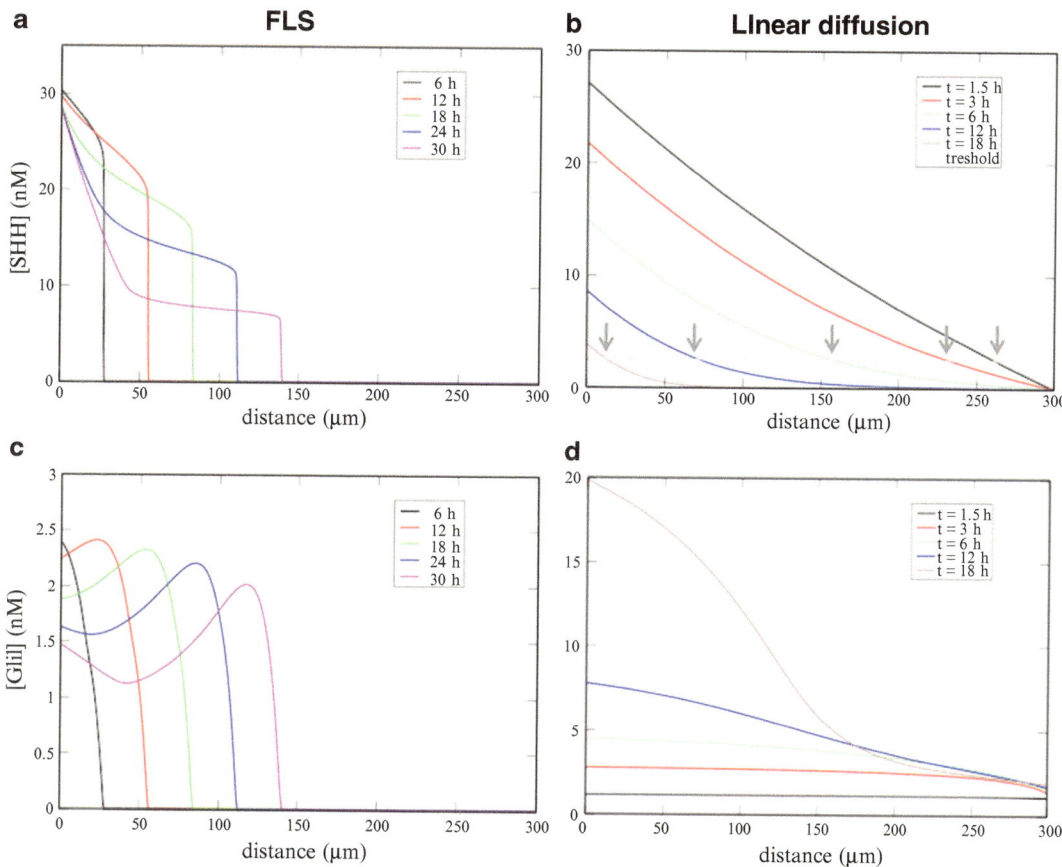

Fig. 4 Plots of Shh, and Gli1[Act] concentrations versus distance from the floor plate at various times. The plots have been obtained numerically solving our FLS (**a, c**) and the linear diffusion model (**b, d**). Note that in linear diffusion modeling (**b, d**) there are no natural fronts and that an artificial threshold had to be imposed at 2.5 nM in order to achieve them (*see* Fig. 8 in ref. [8]), which is independent of any biological reality

similar to the way neurons communicate. This possibility would strengthen the argument that morphogenetic signals do not diffuse freely to form an extracellular concentration gradient [42]. A comparison of results obtained using the FLS model compared to the linear diffusion model shows that it represents the biological complexity with much improved accuracy (Fig. 5). Our thesis is that the biological mechanisms replicated by flux saturation are precisely the cooperation processes between cytonemes and vesicles.

Regarding the macro and micro-descriptions of morphogenetic process and their connection via asymptotic scales, we find that flux-saturated models of biological materials [36, 37, 42] are able to collect the dynamical properties of living matter, or to be extended to aggregates of living beings in the way of swarms, flocks, schools, bacteria colonies, cells, social force, etc. [43–45]. Then flux-saturated mechanism could have the capacity of reproducing some of the emerging behaviors that happen only at the level of the collectivity (i.e., in the vesicles and in its relationship with cytonemes), by going beyond the simplest dynamic of a few entities. Understanding how these molecules gain information from each other, transfer it, cooperate and make decisions is a fascinating issue. Although a broader literature is starting to emerge, the applicability of flux-saturated ideas to this problem was first discussed in ref. [43].

In morphogenesis, the possibility that cell clusters could be connected by cytonemes, transmit information by vesicles and also have the ability to organize their dynamics according to a strategy (based on nonlinear additive actions in the group), constitutes an attractive scientific question that needs to be further explored. The mathematical description of cytonemes requires, at least, a bidimensional version of our model introduced and studied in refs. [37, 38, 41]. Such a model might well incorporate the potential generated by Ptc1–Hh interaction as a source of directionality, and can be a challenging future line of research.

4   Notes

1. We use poly-lysine to prevent imaginal disc movement during imaging, as it will attach the wing imaginal discs to the coverslip.

2. Working at 4 °C will confer a better manipulation of larvae, as they will not move. This condition also prevents medium contamination.

3. Methyl cellulose is used to make the medium more viscous.

4. The promoter term is calculated using statistical thermodynamics and depends on free energies of binding for each

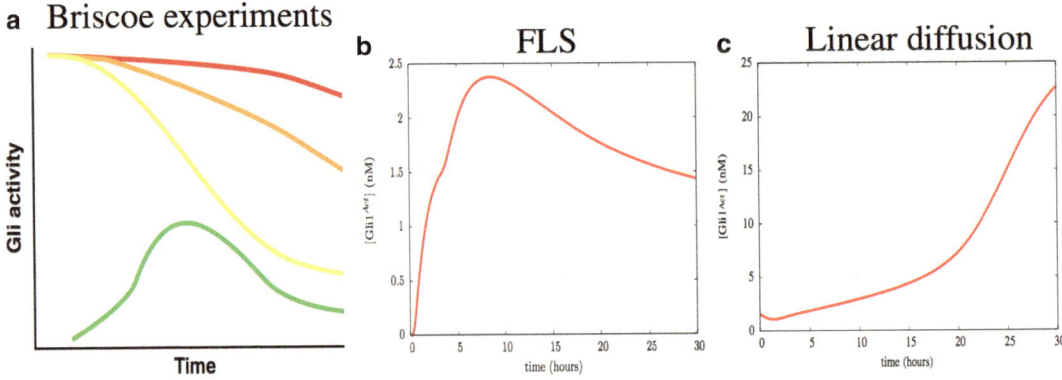

Fig. 5 (**a**) Briscoe's group experiment on desensitization. (**b**, **c**) Evolution of Gli1[Act] over time at a distance of 1 μm from the floor plate resulting from signal accumulation and subsequent desensitization in our FLS model (**b**). This behavior is not observed in the linear diffusion model (**c**). The FLS curve reproduces a temporal adaptation mechanism: after 12 h cells become desensitized to Shh signal and the response decreases

configurations, taking into account the cooperative Gli1 function [26]:

$$P_{tr} = \frac{P_{rel}}{Z_{tot}}$$

where

$$P_{rel} = 3(\in^2 \left[\text{Gli1}^{Act} \right]^3 + \left(\left[\text{Gli3}^{Act} \right] + r \left[\text{Gli1}^{Rep} \right] + K_{Gli3} \right)$$
$$\times \left(\in \left[\text{Gli1}^{Act} \right]^2 + \left[\text{Gli3}^{Act} \right]^2 + \left[\text{Gli1}^{Act} \right] \left(\left[\text{Gli3}^{Act} \right] + r \left[\text{Gli3}^{Rep} \right] + K_{Gli3} \right) \right))$$

and

$$z_{tot} = 3(\in^2 \left[\text{Gli1}^{Act} \right]^3 + 3 \left[\text{Gli3}^{Act} \right]^3 + 3k_{Gli3} \left[\text{Gli3}^{Act} \right]^2 + 3k^2_{Gli3} \left[\text{Gli3}^{Act} \right]$$
$$+ k^3_{Gli3} + 3r \left[\text{Gli3}^{Rep} \right] (K_{Gli3} + \left[\text{Gli3}^{Act} \right]^2$$
$$+ 3r^2 \left[\text{Gli3}^{Rep} \right]^2 \left(K_{Gli3} + \left[\text{Gli3}^{Act} \right] \right)$$
$$+ 3r^3 \left[\text{Gli3}^{Rep} \right]^3 + 3 \in \left[\text{Gli1}^{Act} \right]^2 \left(\left[\text{Gli3}^{Act} \right] + r \left[\text{Gli3}^{Rep} \right] + k_{Gli3} \right)$$
$$+ 3 \left[\text{Gli1}^{Act} \right] \left(\left[\text{Gli3}^{Act} \right] + r \left[\text{Gli3}^{Rep} \right] + K_{Gli3} \right)^2 .$$

Acknowledgements

The paper has been partially supported by Junta de Andalucía Project FQM 954. IG was supported by Fundamental Biology (BFU2011-25987) and Consolider (CDS 2007-00008) program grants from the Spanish Ministry of Economy and Commutativity (MINECO), by Marie Curie FP7- Integration Network (ITN 238186) grant and by an institutional grant to Centro de Biología Molecular "Severo Ochoa" from the Fundación Areces. J.C., O.S., and J.S. were supported in part by Spanish MINECO, project MTM2011-23384 and FEDER funds. JC is also partially supported by La Caixa "Collaborative Mathematical Research" programme and a Juan de la Cierva grant of the spanish MEC.

References

1. Ingham PW, Nakano Y, Seger C (2011) Mechanisms and functions of Hedgehog signalling across the metazoa. Nat Rev Genet 12: 393–406

2. Turing AM (1952) The chemical basis of morphogenesis. Phil Trans Roy Soc Lond Ser B Biol Sci 237:37–72

3. Crick F (1970) Diffusion in embryogenesis. Nature 40:561–563

4. Meinhardt H (1978) Space–dependent cell determination under the control of a morphogen gradient. J Theor Biol 74:307–321

5. Lander AD, Nie Q, Wan FY-M (2002) Do morphogen gradients arise by diffusion? Dev Cell 2:785–796

6. Kondo S, Miura T (2010) Reaction-diffusion model as a framework for understanding biological pattern formation. Science 329:1616–1620

7. Dessaud E, Yang LL, Hill K, Cox B, Ulloa F, Ribeiro A et al (2007) Interpretation of the Sonic Hedgehog morphogen gradient by a temporal adaptation mechanism. Nature 450:717–720

8. Stecca B, Ruiz i Altaba A (2010) Context-dependent regulation of the GLI code in cancer by Hedgehog and non-Hedgehog signals. J Mol Cell Biol 2(2):84–95

9. Jessell TM (2000) Neuronal specification in the spinal cord: inductive signals and transcriptional codes. Nat Rev Genet 1:20–29

10. Guerrero I, Chiang C (2007) A conserved mechanism of Hedgehog gradient formation by lipid modifications. Trends Cell Biol 17:1–5

11. Vyas N, Goswami D, Manonmani A, Sharma P, Ranganath H, VijayRaghavan K, Shashidhara L et al (2008) Nanoscale organization of Hedgehog is essential for long-range signalling. Cell 133:1214–1227

12. Zeng X, Goetz JA, Suber LM, Scott WJ Jr, Schreiner CM, Robbins DJ (2011) A freely diffusible form of Sonic Hedgehog mediates long-range signalling. Nature 411:716–720

13. Greco V, Hannus M, Eaton S (2001) Argosomes: a potential vehicle for the spread of morphogens through epithelia. Cell 106:633–645

14. Panákova D, Sprong H, Marois E, Thiele C, Eaton S (2005) Lipoprotein particles are required for Hedgehog and Wingless signalling. Nature 435:58–65

15. Callejo A, Culi J, Guerrero I (2008) Patched, the receptor of Hedgehog, is a lipoprotein receptor. Proc Natl Acad Sci U S A 105:912–917

16. Gurdon JB, Mitchell A, Mahony D (1995) Direct and continuous assessment by cells of their position in a morphogen gradient. Nature 376:520–521

17. Saha K, Schaffer DV (2006) Signaling dynamics in Sonic hedgehog tissue patterning. Development 133:889–900

18. Kornberg TB (2012) The imperatives of context and contour for morphogen dispersion. Biophys J 103:2252–2256

19. Ramirez-Weber FA, Kornberg TB (1999) Cytonemes: cellular processes that project to the principal signaling center in Drosophila imaginal discs. Cell 97:599–607

20. Bilioni A, Sánchez-Hernández D, Callejo A, Gradilla AC, Ibáñez C, Mollica E et al (2013) Balancing Hedgehog, a retention and release equilibrium given by Dally, Ihog, Boi and shifted/DmWif. Dev Biol 376:198–212

21. Callejo A, Bilioni A, Mollica E, Gorfinkiel N, Andrés G, Ibáñez C et al (2011) Dispatched mediates Hedgehog basolateral release to form the long-range morphogenetic gradient in the Drosophila wing disk epithelium. Proc Natl Acad Sci U S A 108:12591–12598

22. Gradilla A-C, González E, Seijo I, Andrés G, González-Méndez L, Sánchez V et al (2014) Exosomes as Hedgehog carriers in cytoneme-mediated transport and secretion. Nat Commun 5:5649

23. Roy S, Hsiung F, Kornberg TB (2011) Specificity of Drosophila cytonemes for distinct signalling pathways. Science 15:354–358

24. Sanders TA, Llagostera E, Barna M (2013) Specialized filopodia direct long-range transport of Shh during vertebrate tissue patterning. Nature 497:628–632

25. Bischoff M, Gradilla AC, Seijo I, Andrés G, Rodríguez-Navas C, González-Méndez L, Guerrero I (2013) Cytonemes are required for the establishment of a normal Hedgehog morphogen gradient in Drosophila epithelia. Nat Cell Biol 15:1269–1283

26. Verbeni M, Sánchez O, Mollica E, Siegl-Cachedernier I, Carleton A, Guerrero I et al (2013) Morphogenetic action through flux-limited spreading. Phys Life Rev 10:457–475

27. Zheng X, Mann RK, Sever N, Beachy PA (2010) Genetic and biochemical definition of the Hedgehog receptor. Genes Dev 24:57–71

28. Izzi L, Lévesque M, Morin S, Laniel D, Wilkes BC, Mille F et al (2011) Boc and Gas1 each form distinct Shh receptor complexes with Ptch1 and are required for Shh-mediated cell proliferation. Dev Cell 20:788–801

29. Yan D, Wu Y, Yang Y, Belenkaya TY, Tang X, Lin X et al (2010) The cell-surface proteins Dally-like and Ihog differentially regulate Hedgehog signaling strength and range during development. Development 137:2033–2044

30. Cardozo MJ, Sánchez-Arrones L, Sandonis A, Sánchez-Camacho C, Gestri G, Wilson SW, Guerrero IP (2014) Bovolenta, Cdon acts as a Hedgehog decoy receptor during proximal-distal patterning of the optic vesicle. Nat Commun 5:4272

31. Brand AH, Perrimon N (1991) Generating lineage-specific markers to study Drosophila development. Dev Genet 12:238–252

32. Torroja C, Gornkiel N, Guerrero I (2004) Patched controls the Hedgehog gradient by endocytosis in a dynamic-dependent manner, but this internalization does not play a major role in signal transduction. Development 131:2395–2408

33. Tanimoto H, Itoh S, ten Dijke P, Tabata T (2000) Hedgehog creates a gradient of DPP activity in Drosophila wing imaginal discs. Mol Cell 5:59–71

34. Ashburner M, Roote J (2007) Maintenance of a Drosophila laboratory: general procedures. CSH Protoc 2007: pdb.ip35

35. Rojas-Rios P, Guerrero I, Gonzalez-Reyes A (2012) Cytoneme-mediated delivery of hedgehog regulates the expression of bone morphogenetic proteins to maintain germline stem cells in Drosophila. PLoS Biol 10: e1001298

36. Bellomo N, Bellouquid A, Nieto J, Soler J (2010) Multiscale biological tissue models and flux-limited chemotaxis for multicellular growing systems. Math Models Methods Appl Sci 20:1179–1207

37. Campos J, Guerrero P, Sánchez O, Soler J (2013) On the analysis of travelling waves to a nonlinear flux limited reaction-diffusion equation. Ann Inst H Poincaré Anal Non Linéaire 30:141–155

38. Calvo J, Campos J, Caselles V, Sánchez O, Soler J. Pattern formation in a flux limited reaction-diffusion equation of porous media type. http://arxiv.org/abs/1309.6789

39. Rosenau P (1992) Tempered diffusion: a transport process with propagating front and inertial delay. Phys Rev A 46:7371–7374

40. Brenier Y (2003) Extended Monge-Kantorovich theory. In: Caffarelli LA, Salsa S (Eds) Optimal transportation and applications, Lectures given at the C.I.M.E. Summer School help in Martina Franca, Lecture Notes in Math. 1813, Springer-Verlag. pp 91–122

41. Calvo J, Mazón JM, Soler J, Verbeni M (2011) Qualitative properties of the solutions of a nonlinear flux–limited equation arising in the transport of morphogens. Math Models Methods Appl Sci 21:893–937

42. Guerrero I, Kornberg TB (2014) Hedgehog and its circuitous journey from producing to target cells. Semin Cell Dev Biol 33C:52–62

43. Bellomo N, Soler J (2012) On the mathematical theory of the dynamics of swarms viewed as a complex system. Math Models Methods Appl Sci 22, Paper No. 1140006

44. Ballerini M, Cabibbo N, Candelier R, Cavagna A, Cisbani E, Giardina I et al (2008) Interaction ruling animal collective behavior depends on topological rather than metric distance: evidence from a field study. Proc Natl Acad Sci U S A 105(4):1232–1237

45. Couzin ID (2007) Collective minds. Nature 445:715

Evaluating the Activity of Smoothened Toward G Proteins Using [^{35}S]Guanosine 5′-(3-*O*-thio)triphosphate ([^{35}S]GTPγS)

David R. Manning, Feng Shen, and Natalia A. Riobo

Abstract

The utilization of heterotrimeric G protein, and in particular those of the G$_i$, family, by Hedgehogs through Smoothened has become increasingly clear. We describe here a method for evaluating the activity of Smoothened toward G proteins in membranes derived from human embryonic kidney-293 (HEK293) cells. The assay relies on receptor-promoted exchange of GDP for [^{35}S]GTPγS on the Gα subunit. The assay is best suited for analysis of the constitutive activity of Smoothened, inverse agonism superimposed on this activity, and neutral antagonism superimposed on inverse agonism. The assay would also be suitable for several other applications requiring a proximal, quantifiable readout of Smoothened activity.

Key words Hedgehog, Smoothened, G protein, G$_i$, GTPγS, HEK293

1 Introduction

Smoothened (Smo) has long been posited to act in part through heterotrimeric GTP-binding regulatory proteins (G proteins) based on its 7-transmembrane helical motif and sensitivity of some of its actions to a *B. pertussis* toxin acting overtly on the G$_i$ family of G proteins [1, 2]. The first direct evidence for Smo coupling to G proteins, however, came with the demonstration that Smo promotes binding of [^{35}S]guanosine 5′-(3-*O*-thio)triphosphate ([^{35}S] GTPγS) to α subunits of individual forms of G$_i$ [3]. Techniques utilizing [^{35}S]GTPγS exploit a property fundamental to receptor·G-protein communication—an activated receptor catalyzes exchange of GDP for GTP (here, [^{35}S]GTPγS) as a means of activating the G protein. [^{35}S]GTPγS is an analogue of GTP that can be obtained with high specific activity and does not easily undergo hydrolysis. Assays of [^{35}S]GTPγS binding are routinely linked to immunoprecipitation with G-protein α subunit-selective antibodies

Natalia A. Riobo (ed.), *Hedgehog Signaling Protocols*, Methods in Molecular Biology, vol. 1322, DOI 10.1007/978-1-4939-2772-2_4, © Springer Science+Business Media New York 2015

under nondenaturing conditions in order to define precisely the G protein activated.

Evaluation of G-protein activation using [35S]GTPγS can be carried out for Smo expressed in any number of cells [3–5]. We detail here a protocol devised for evaluating Smo in human embryonic kidney-293 (HEK293) cells [4, 5]. HEK293 cells are a widely used platform for overexpression and analysis of receptors coupled to G proteins. Smo expresses a particularly high level of constitutive activity in an overexpression paradigm, probably attributable to an amount of Smo in excess of endogenous Ptch1. Indeed, agonist-dependent [35S]GTPγS binding has so far been detected only in the instance of Smo expressed endogenously in mouse embryonic fibroblasts [3] and, to a very small extent, as a construct complexed to $G\alpha_{i2}$ [5] (*see* below). Overexpression of Smo in HEK293 cells lends itself best, therefore, to evaluation of the constitutive activity of Smo, of inverse agonism superimposed on this activity, and of neutral antagonism superimposed on inverse agonism. The HEK293 cell setting with [35S]GTPγS binding as the readout can conceivably be extended to evaluate structure/activity relationships intrinsic to Smo and Smo's interactions with other (co-introduced) proteins in so far as the latter proteins affect activity. In this sense, the activation of one or more forms of G_i as monitored by [35S]GTPγS binding represents one of the most proximal measures of Smo activity available.

2 Materials

2.1 Buffers

1. HE with protease inhibitors: 20-mM Hepes, pH 8.0, 1-mM EDTA, 10-μg/ml leupeptin, 10-μg/ml aprotinin, and 100-μM PMSF (*see* **Note 1**).

2. TMEN: 50-mM Tris–HCl, pH 7.4, 20-mM $MgCl_2$, 2-mM EDTA, and 100-mM NaCl.

3. TMEN/GDP: TMEN containing 10-μM GDP (*see* **Note 2**).

4. TMN: 50-mM Tris–HCl, pH 7.5, 20-mM $MgCl_2$, 150-mM NaCl.

5. TMN/NP-40: TMN containing 0.5 % (v/v) Nonidet P-40.

6. TMN/NP-40/GDP/GTP/aprotinin: TMN/NP-40 containing 100-μM GDP, 100-μM GTP, and 0.33 % aprotinin.

2.2 Cell Culture Reagents

1. Human embryonic kidney (HEK293) cells (American Tissue Culture Collection, Manassas, VA).

2. Dulbecco's modified Eagle's medium (DMEM).

3. Fetal bovine serum.

4. Penicillin/streptomycin solution (100×).

5. Phosphate-buffered saline.

6. Lipofectamine 2000 (Gibco/Life Sciences, Grand Island, NY).

7. pcDNA3.1-based plasmids or similar encoding Smo or mutants.

2.3 Other Reagents or Materials

1. [^{35}S]GTPγS (PerkinElmer Life and Analytical Sciences, Boston, MA). The nominal specific activity of the [^{35}S]GTPγS is 1,250 Ci/mmol, and the concentration is approximately 10 μM. 4.96 μl of the purchased [^{35}S]GTPγS is diluted into 395-μl TMEN to make a stock solution of 124-nM [^{35}S]GTPγS (*see* **Note 3**). The stock is stored at –80 ° C. The [^{35}S]GTPγS is stable for several weeks, though specific radioactivity will decrease. It can be thawed and refrozen many times (*see* **Note 4**).

2. Antisera/antibodies: We use peptide-directed rabbit antisera raised toward to the C termini of $G\alpha_{i1}$, $G\alpha_{i2}$, and $G\alpha_{i3}$ (*see* **Note 5**). We make the antisera ourselves out of convenience [6]. Commercially available purified antibodies that are similarly directed, for example, some of those from Santa Cruz Biotechnology (Dallas, TX), are compatible with the protocol (*see* **Note 6**).

3. Protein A immobilized on Sepharose CL-4B beads, pre-equilibrated (Sigma-Aldrich, St. Louis, MO). We typically swell 0.5 g of the beads in 10-ml 10-mM HEPES, pH 7.4, for 30 min at 4 °C and then pellet them at $4,000 \times g$ for 5 min. The beads are resuspended in 10-ml TMN/NP-40 containing 4-mg/ml bovine serum albumin and shaken 30 min at 4 °C. The beads are again pelleted and then washed twice through resuspension and pelleting with TMN/NP-40 without bovine serum albumin. The beads are resuspended and stored at 4 °C in 10-ml TMN/NP-40 containing 0.33 % (v/v) aprotinin.

4. Pansorbin® cells (EMD Millipore, Billerica, MA).

5. Ecolite(+)™ liquid scintillation cocktail (MP Biomedicals, Santa Ana, CA).

6. BCA protein assay kit (Thermo Scientific, Rockford, IL).

2.4 Equipment

1. Humidified incubator at 37 °C with 5 % CO_2.

2. Refrigerated microcentrifuge, such as Eppendorf 5414R.

3. Liquid scintillation counter.

4. Orbital rotator.

5. Water bath at 30 °C.

3 Methods

The following procedure describes the processing of twenty 10-cm plates of confluent HEK293 cells. Cells from this number of plates provide about 600 μg of membrane protein, enough for 30 assay points or ten "conditions" in triplicate. Conditions include membranes from (1) cells transfected with vector that does not express Smo, as a control for specificity of the assay for Smo action, (2) cells transfected with vector expressing Smo, to define constitutive activity, and (3) cells transfected with vector expressing Smo then incubated with agonist, inverse agonist, and/or neutral antagonist at varying concentrations, to evaluate ligand impact on constitutive activity and/or the actions of each other.

3.1 Transfection and Preparation of Membranes from HEK293 Cells

1. Plate HEK293 cells in 10-cm tissue culture dishes in DMEM with 10 % fetal bovine serum at 37 °C with 5 % CO_2.

2. Transfect HEK293 cells when the cells are 70–80 % confluent (*see* **Notes 7–9**). Lipofectamine 2000 is used according to the manufacturer's instructions. Typically, we use 40 µg of pcDNA-Smo per 10-cm dish. Some cells, for example, 2 or 3 plates, should be transfected with vector that does not express Smo, i.e., pcDNA3.1, to determine through comparison the signal specifically attributable to the Smo that was introduced. Membranes (below) are prepared about 48 h following transfection, a time at which the cells have reached confluence (*see* **Note 10**).

3. At about 48 h after transfection, rinse the transfected HEK293 cells in monolayer with phosphate-buffered saline. Remove the phosphate-buffered saline and add to each plate 0.5-ml ice-cold HE buffer with protease inhibitors. Rock the plates for 20 min at 4 °C.

4. For each plate, scrape the cells and transfer them using a 1-ml PIPETMAN to a 1.5-ml microfuge tube on ice. Using a pre-cooled 1-ml plastic syringe, pass the cells up and down 15 times through a 26G needle to achieve cell lysis.

5. Centrifuge the lysates in the microfuge tubes at $650 \times g$ for 5 min at 4 °C to remove nuclei and unlysed cells (*see* **Note 11**).

6. Transfer the supernatants to a new set of microfuge tubes and recentrifuge at $20,000 \times g$ for 30 min at 4 °C to pellet cell membrane.

7. Discard the supernatants and resuspend each pellet in 100 µl of HE with protease inhibitors, again at 4 °C using a syringe fitted with a 26G needle. Combine resuspended membrane from, respectively, cells transfected with empty vector and cells transfected with vector expressing Smo.

8. Take a small portion of resuspended membranes for BCA protein assay. The remainder is used for the $[^{35}S]GTP\gamma S$ assay (*see* **Note 12**).

3.2 Prepare Pansorbin® Cells for Preclearing

1. Add 100 µl of evenly suspended Pansorbin® cells as provided by the manufacturer to each of 24 1.5-ml microfuge tubes (*see* **Notes 13** and **14**). Centrifuge at $10,000 \times g$ for 1 min at 4° and discard the supernatant.

2. Add 2 μl of normal rabbit serum (i.e., preimmune or "nonimmune" serum) followed by 25 μl of TMN to each of the Pansorbin® cell pellets and resuspend.

3. Shake the tubes for at least 2 h at 4 °C in orbital shaker (*see* **Note 15**).

3.3 Prepare Antisera/Antibodies

1. Gα-directed antibodies are prebound to protein A-Sepharose at the same time that Pansorbin® cells are prepared (Subheading 3.2). Add 10 μl of antiserum to 100 μl of evenly suspended protein A-Sepharose beads in each of 24 1.5-ml microfuge tubes.

2. Shake the tubes for at least 2 h at 4 °C in orbital shaker (*see* **Note 15**).

3.4 [^{35}S]GTPγS Binding

1. Prepare a 1:10 dilution of the [^{35}S]GTPγS stock in TMEN to achieve 12.4-nM [^{35}S]GTPγS.

2. Dilute the ligand of interest into H_2O, or other vehicle, to a concentration 31-fold that is desired in the assay.

3. Centrifuge the membrane suspensions prepared in Subheading 3.1 at 20,000×g for 30 min at 4 °C. Discard the supernatants and resuspend the membranes in TMEN/GDP at a protein concentration of 0.36 μg/μl using 26G needles fitted to 1-ml syringes. Store on ice until needed.

4. Pipette 55 μl of the membrane suspensions (20-μg membrane protein) into each of 24 1.5-ml microfuge tubes. Generally, membranes from HEK293 cells transfected with vector alone will be dispensed into 3–6 microfuge tubes, and membranes from HEK293 cells transfected with vector expressing Smo will be dispensed into the remaining tubes.

5. Add 2 μl of ligand or vehicle to the membrane in each of the microfuge tubes, vortex briefly, and incubate at 30 °C for 10 min (*see* **Note 16**).

6. Add 5 μl of the diluted (12.4-nM) [^{35}S]GTPγS to each of the tubes, vortex briefly, and incubate at 30 °C for an additional 10 min (*see* **Note 17**).

7. Stop the reaction by removing the tubes to ice and adding 600 μl of ice-cold TMN/NP-40/GDP/GTP/aprotinin to each tube. Shake the tubes for 30 min at 4 °C in orbital shaker (*see* **Note 15**).

3.5 Immuno-precipitation

1. Transfer the contents of each reaction tube in Subheading 3.4 to a microfuge tube containing Pansorbin cells and irrelevant antibodies (*see* Subheading 3.2). Shake for 30 min at 4 °C in orbital shaker to eliminate nonspecific binding (*see* **Note 15**).

2. Centrifuge the tubes for 3 min at 20,800×g at 4 °C.

3. Transfer the supernatants to the microfuge tubes containing antibodies prebound to protein A-Sepharose (*see* Subheading 3.3) and shake for 1 h at 4 °C in orbital shaker to allow binding of solubilized Gα subunits to the Sepharose-coupled antibody (*see* **Note 15**).

4. Pellet the beads by centrifugation at $20{,}800 \times g$ for 3 min at 4 °C.

5. Discard the supernatant and resuspend the beads in 0.5-μl TMN/NP-40/GDP/GTP/aprotinin. Vortex briefly and collect the beads by centrifugation as above and wash two more times by resuspension and pelleting (*see* **Note 18**). Wash an additional time in TMN alone, centrifuge and discard the supernatant.

6. Add 500 μl of 0.5 % SDS to each sample. Place the tubes containing each SDS-treated sample in a boiling water bath for 1 min, and then transfer the entire content into a scintillation vial. Add 5-ml scintillation fluid and submit to scintillation spectrometry.

4   Notes

1. Add protease inhibitors to the buffer just before using. PMSF in particular is unstable in water.

2. Inclusion of 10-μM GDP in the assay reduces receptor-independent binding of $[^{35}S]GTP\gamma S$ [7].

3. ^{35}S emits β particles as it decomposes. At this and all other steps involving $[^{35}S]GTP\gamma S$, including the storage and disposal of $[^{35}S]GTP\gamma S$, be sure to use appropriate precautions to minimize exposure to radioactivity. Water and plexiglass are particularly effective barriers.

4. We note a falloff in the quality of data after about 2 months of storage of $[^{35}S]GTP\gamma S$. The falloff extends beyond simple decay of radioactivity. We suspect it is decomposition of some sort, quite possibly oxidation.

5. We do not bother to purify antibodies from antisera, as it makes no difference to the results. Antibodies directed to portions of a Gα subunit apart from the C terminus may or may not work. The chief consideration is an ability to immunoprecipitate efficiently under nondenaturing conditions and without otherwise disrupting the $G\alpha \cdot [^{35}S]GTP\gamma S$ complex. The C termini of $G\alpha_{i1}$ and $G\alpha_{i2}$ are identical. Generally, antibodies are directed to these two subunits specifically. The C terminus of $G\alpha_{i3}$ is similar enough to those of $G\alpha_{i1}$ and $G\alpha_{i2}$ to exhibit substantial cross-reactivity.

6. Antisera and antibodies may differ considerably in avidity, even over time from a single source. We suggest in our protocol 10 μl of antisera for each assay point. If using purified antibody, we suggest 1 μg of purified antibody as a starting point in optimization.

7. HEK293 cells contain only very small amounts of endogenous Smo. The amounts are too small to be detected by the assay [5].

8. A great deal of attention is paid to the degree of cell confluence in studies of Hedgehog signaling owing to contact-dependent formation of primary cilia. Our studies are always conducted in membranes prepared from confluent cells, but we have not evaluated whether confluency is truly required for successful assay of G-protein activation.

9. On occasion, we transfect cells with vector encoding a Smo·Gα fusion protein, which consists of Smo truncated to near the C-terminal aspect of its seventh transmembrane domain (as the C tail of Smo is very long) and the Gα subunit of interest [5]. This generally permits amplification of [^{35}S]GTPγS binding without loss of specificity. It also permits, using Gα-selective antibodies and Western blots, normalization of fusion protein expression for comparison among different receptor fusion constructs (Fig. 1).

10. In experiments with HA-tagged Smo, we find, using flow cytometry, that about 50 % of HEK293 cells express Smo at the cell surface 48 h after transfection (Fig. 2) [5].

11. We use an Eppendorf 5417C centrifuge placed in a cold room for this purpose.

12. We generally proceed directly into the [^{35}S]GTPγS-binding assay from the protein assay in the same day. If need be, the membrane can be quick-frozen in a dry ice/methanol mixture and stored at −80 °C for future assay.

13. Pansorbin cells treated with nonimmune sera in this section are used in later steps (see Subheading 3.5, "Immunoprecipitation") to remove proteins in solubilized membranes that adhere non-specifically to protein A and/or immunoglobulins.

14. The number of tubes is limited by the amount that your micro-centrifuge rotor can accommodate, as centrifugation of a set of tubes followed by centrifugation of another set will give anomalous results.

15. The Eppendorf tubes must be securely closed and shaken vigorously horizontally. Shaker speed is critical; we use 180 rpm. To facilitate the assembly, we place the tubes in a microtube rack with lid and tape it into a container attached to the orbital shaker.

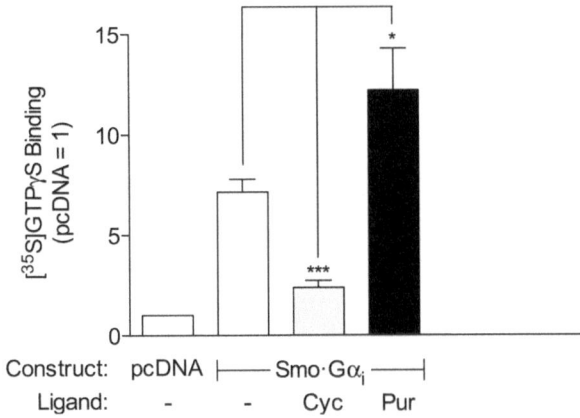

Fig. 1 Activation of $G\alpha_i$ within a $Smo \cdot G\alpha_i$ fusion protein. HEK293 cells were transfected with pcDNA3.1 alone or the vector encoding the $Smo \cdot G\alpha_{i2}$ fusion protein. $[^{35}S]GTP\gamma S$ binding was evaluated with or without the Smo agonist purmorphamine (10 μM) or the Smo inverse agonist cyclopamine (10 μM) and normalized to binding obtained with pcDNA3.1 alone. Differences of $Smo \cdot G\alpha_{i2}$ with and without ligands are noted ($*p \leq 0.05$; $***p \leq 0.001$). Note the activity exhibited upon overexpression of the fusion protein $Smo \cdot G\alpha_{i2}$ versus pcDNA without ligand (constitutive activity), the suppression of that activity by the inverse agonist cyclopamine, and the enhancement of that activity by the agonist purmorphamine. The $Smo \cdot G\alpha_{i2}$ fusion construct represents an infrequent instance of overexpressed Smo (in this case as a fusion protein) exhibiting sensitivity to an agonist, i.e., purmorphamine. Reproduced from Feng et al. 2013, with modification by permission from the American Society for Pharmacology and Experimental Therapeutics [5]

16. Having ligand and receptor come to equilibrium simplifies the analysis.

17. We find 10 min for binding of $[^{35}S]GTP\gamma S$ to be optimal; however, in careful comparisons among ligands, it is important to establish linearity in $[^{35}S]GTP\gamma S$ binding as a function of time, for example, to evaluate binding over a range of 2–20 min.

18. After binding and centrifugation, we observe a discrete white precipitate right above the upper portion of the angled pelleted beads. This white "dot" must be aspirated and discarded together with the supernatant since it contains a great amount of nonbound radioactivity. After the last wash, the "dot" must not be visible any more.

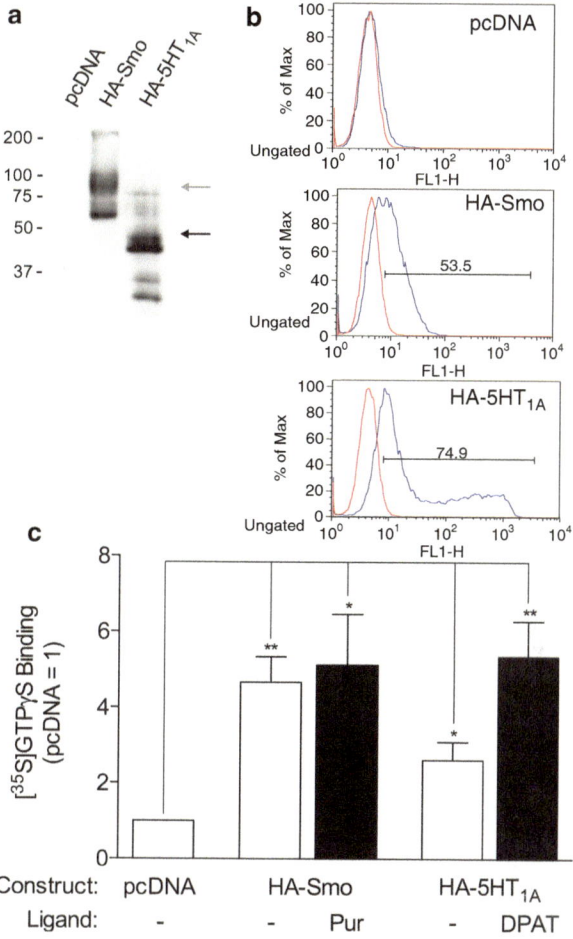

Fig. 2 Activation of G_i endogenous to HEK293 cells by Smo. HEK293 cells near confluence were transfected with pcDNA3.1 alone or the vector encoding HA-tagged Smo or, for comparison, HA-tagged 5-HT$_{1A}$ receptor. Cells or subsequently prepared membranes were evaluated 48 h following transfection. (**a**) Expression of the receptors was evaluated for membranes (1-μg membrane protein) by a Western blot using HA-directed antibody. Positions of molecular weight standards are noted. *Arrows* refer to mobilities predicted for full-length receptors (*gray arrow*, HA-tagged Smo; *black arrow*, HA-tagged 5-HT$_{1A}$ receptor) based on molecular weight standards. The depiction is representative of four experiments in total. (**b**) Expression of receptors was evaluated for intact cells by flow cytometry using either no primary antibody (*red*) or HA-directed antibody (*blue*) with FITC-conjugated secondary antibody. *Horizontal bars* represent cutoffs for expression and equate to transfection efficiency. The depiction is representative of three experiments in total. (**c**) Activity of the receptors toward one or more forms of G_i endogenous to HEK293 cells was evaluated in membranes by [^{35}S]GTPγS binding. Binding of the radioligand was examined with or without 10-μM purmorphamine (HA-Smo) or 1-μM 8-OH-DPAT (HA-5HT$_{1A}$) and normalized to binding obtained with pcDNA alone. Differences from pcDNA3.1 alone are noted (*$p \leq 0.05$; **$p \leq 0.01$). Reproduced from Feng et al. 2013, by permission from the American Society for Pharmacology and Experimental Therapeutics [5]

Acknowledgments

Our work on the use of G proteins by Smoothened is supported by the National Institutes of Health grant GM080396.

References

1. DeCamp DL, Thompson TM, de Sauvage FJ, Lerner M (2000) Smoothened activates $G\alpha_i$-mediated signaling in frog melanophores. J Biol Chem 275:26322–26327

2. Kanda S, Mochizuki Y, Suematsu T, Miyata Y, Nomata K, Kanetake H (2003) Sonic hedgehog induces capillary morphogenesis by endothelial cells through phosphoinositide 3-kinase. J Biol Chem 278:8244–8249

3. Riobo NA, Saucy B, DiLizio C, Manning DR (2006) Activation of heterotrimeric G proteins by Smoothened. Proc Natl Acad Sci U S A 103:12607–12612

4. Douglas AE, Heim JA, Shen F, Almada LL, Riobo NA et al (2011) The α subunit of the G protein G_{13} regulates activity of one or more Gli transcription factors independently of Smoothened. J Biol Chem 286:30714–30722

5. Shen F, Cheng L, Douglas AE, Riobo NA, Manning DR (2013) Smoothened is a fully competent activator of the heterotrimeric G protein G_i. Mol Pharmacol 83:691–697

6. Butkerait P, Zheng Y, Hallak H, Graham TE, Miller HA, Burris KD et al (1995) Expression of the human 5-hydroxytryptamine$_{1A}$ receptor in Sf9 cells: Reconstitution of a coupled phenotype by co-expression of mammalian G protein subunits. J Biol Chem 270:18691–18699

7. Windh RT, Manning DR (2002) Analysis of G protein activation in Sf9 and mammalian cells by agonist-promoted [35S] GTPγS binding. Methods Enzymol 344:3–14

Chapter 5

Analysis of Smoothened Phosphorylation and Activation in Cultured Cells and Wing Discs of *Drosophila*

Kai Jiang and Jianhang Jia

Abstract

Smoothened (Smo) is essential for transduction of the Hedgehog (Hh) signal in both insects and vertebrates. Binding of Hh to Ptc-Ihog relieves the Patched (Ptc)-mediated inhibition of Smo, which allows Smo to activate the cubitus interruptus (Ci)/Gli family of zinc finger transcription factors and thereby induce the expression of Hh target genes, such as *decapentaplegic* (*dpp*), *ptc*, and *engrailed* (*en*). The activation of Smo appears to be one of the most important events in Hh signaling. Studies have shown that Hh induces cell surface/ciliary accumulation and phosphorylation of Smo by multiple kinases, including protein kinase A (PKA), casein kinase 1 (CK1), casein kinase 2 (CK2), G protein-coupled receptor kinase 2 (Gprk2), and atypical PKC (aPKC). Here, we describe the assays used to examine the activity of Smo in Hh signaling, including in vitro kinase, *ptc*-luciferase reporter assay, cell surface accumulation assay, fluorescence resonance energy transfer (FRET) assay, and wing disc immunostaining. These assays are powerful tools to study Smo phosphorylation and activation, which have provided mechanistic insight into a better understanding the mechanisms of Smo regulation.

Key words Smoothened, Phosphorylation, Hedgehog, Cell surface accumulation, Anti-SmoP antibody, Wing disc immunostaining

1 Introduction

Hedgehog (*hh*) was originally discovered as a segment polarity gene involved in *Drosophila* embryo development [1]. It has been shown that Hh family members function as morphogens and play critical roles in pattern formation and cell growth control; therefore, aberrant Hh signaling causes birth defects as well as several types of cancer [2–4]. The Hh signal is transduced by a signaling cascade highly conserved among different species. Smoothened (Smo), an atypical G protein-coupled receptor (GPCR), is essential for transduction of the Hh signal in both insects and vertebrates. Abnormal activation of Smo results in basal cell carcinoma and medulloblastoma; thus, Smo is an attractive therapeutic target. Currently, the most advanced drug is vismodegib that was approved by FDA for

Natalia A. Riobo (ed.), *Hedgehog Signaling Protocols*, Methods in Molecular Biology, vol. 1322,
DOI 10.1007/978-1-4939-2772-2_5, © Springer Science+Business Media New York 2015

treating cancers driven by Smo activation. However, cancer cells can acquire resistance through mutations in Smo, and vismodegib becomes ineffective. Therefore, a better understanding of the mechanisms of Smo regulation in the fundamental development processes is critical to developing more effective therapeutic treatments for cancers caused by Smo dysregulation.

Studies have shown that Hh induces cell surface accumulation and phosphorylation of Smo by multiple kinases, including protein kinase A (PKA) and casein kinase 1 (CK1), casein kinase 2 (CK2), G protein-coupled receptor kinase 2 (Gprk2), and atypical PKC (aPKC) which activate Smo by inducing differential phosphorylation, thus the conformational change in the protein [5–12]. In mammals, Hh signal transduction depends on the primary cilium, and ciliary accumulation is required for Smo activation [13–16]. Moreover, phosphorylation by multiple kinases promotes the ciliary localization of mammalian Smo [17]. Therefore, the primary cilium may function as a signaling center for the Hh pathway in mammals [18]. Smo C-tail phosphorylation by PKA and CK1 has been derived mainly from a combination of in vitro kinase assay and in vivo functional assay of phosphorylation site mutations. Direct determination of individual site phosphorylation of Smo in vivo is important to the field. After generating an anti-SmoP antibody to specifically detect phosphorylation at the second cluster, regulation of Smo phosphorylation by different levels of Hh signaling activity has been studied [11]. Meanwhile, the luciferase reporter assay is widely used as a tool to study gene expression at the transcriptional level, because it is convenient, relatively inexpensive, and gives quantitative measurements instantaneously.

One of the best model systems to study Hh signal transduction is the *Drosophila* wing imaginal disc that is divided into posterior (P) and anterior (A) compartments. The P compartment cells express and secrete Hh proteins that act upon neighboring A compartment cells located adjacent to the A/P boundary to induce the expression of *dpp* [19, 20]. The Dpp protein then diffuses bidirectionally into both the A and P compartments and functions as a morphogen to control the growth and patterning of cells in the entire wing in a concentration-dependent manner [21–23]. Hh also activates other genes, including *col*, *ptc*, and *en*. Low levels of Hh are able to induce the expression of *dpp*, whereas higher levels of Hh are also able to activate *col* and *ptc*. The induction of *en* appears to require the highest doses of Hh signaling activities [24, 25]. In this chapter, we describe how to use these methods to characterize Smo phosphorylation and activation in cultured cells and wing discs of *Drosophila*.

A new technology to facilitate in vivo genetic manipulations of *Drosophila* with ΦC31 integrase-mediated transgenesis has been recently developed [26, 27]. This strategy causes a clean exchange of the target sequence cassette (attP) with the donor cassette (attB).

Recombination will produce two new sequences, attR and attL, which are not recognized by the integrase, thus generating stable integrants that cannot be excised. Some attP docking lines (VK lines) with *piggyBac-y+-attP* docking element transformed in the *Drosophila* genome have been characterized [27]. In combination with the optimized ΦC31 integrase-expressing strain [28], a *vas-phi-zh2A-VK5* fly stock containing ΦC31 integrase on the X chromosome and the attP site (VK5) on the third chromosome was generated (gift from Dr. Hugo Bellen, Baylor College of Medicine). We have been using this VK5 attP site to examine the in vivo activity of different forms of Smo, either for their ability to rescue *smo³* null mutant or for their ectopic activity when expressed in wing imaginal disc [9, 11, 12, 29]. This approach has allowed us to determine the precise signaling activity of individual Smo variants.

In this chapter, we discuss several methodologies to study the activity and phosphorylation status of Smo in *Drosophila*. We will detail how to culture and transfect S2 cells, evaluate activation of a *ptc*-luciferase reporter, perform in vitro and in vivo phosphorylation assays with a specific Smo construct but which can be adapted to different constructs and mutations, and, finally, generate knock-in transgenic flies harboring Smo phospho-site mutants.

2   Materials

All solutions are prepared using ultrapure water (purified by Milli-Q integral water purification system). All chemicals are analytical-grade or molecular biology-grade reagents.

2.1   Cell Culture and Transfection

1. *Drosophila* Schneider 2 cells, abbreviated as S2 cells, are one of the most commonly used *Drosophila* cell lines. S2 cells were derived from a primary culture of late-stage *Drosophila* embryos and are commercialized by several vendors.

2. S2 cell culture medium: Schneider's *Drosophila* Medium (Invitrogen) with 10 % fetal bovine serum (Sigma), 100 U/mL penicillin, and 100 µg/mL streptomycin (Invitrogen).

3. Hh-conditioned medium: obtained from Hh stable cell line of S2 cells after 24-h induction with 0.7 mM $CuSO_4$.

4. Effectene transfection reagent (QIAGEN).

5. Transfection-grade plasmids: Ub-Gal4 and desired expression constructs cloned into pUAST vectors.

6. Cell culture incubator at 25 °C.

2.2   Luciferase Reporter Assay

1. Transfection-grade plasmids: *ptc*-luciferase and Renilla reporter constructs, Ub-Gal4, tub-Ci (Ci under the α-tubulin promoter), and the desired Smo constructs.

2. Firefly (*Photinus pyralis*) and Renilla (*Renilla reniformis* or sea pansy) luciferase detection kit, such as Dual-Luciferase Reporter Assay System (Promega).

3. GloMax Multi Detection System Luminometer (Promega).

4. Rocker.

2.3 Anti-SmoP Antibody Production

The anti-SmoP antibody, which specifically detects Smo phosphorylation at the second phosphorylation cluster, was generated by Genemed Synthesis Inc. by injecting the antigen peptide CRHVSVESRRN(pS)VD(pS)QV(pS)VK into rabbits. The serum was affinity purified with the antigen, and the flow through was kept as a control antibody against non-phosphorylated peptide.

2.4 GST-Smo Fusion Protein Expression and Purification

1. *Escherichia coli* BL21 (DE3) strain.

2. *E. coli* culture medium: Luria-Bertani (LB) medium with 100 μg/mL ampicillin. Add 0.1 mM isopropyl-beta-D-thiogalactopyranoside (IPTG) for inducing fusion protein expression.

3. Phosphate-buffered saline (PBS): 140 mM NaCl, 2.7 mM KCl, 10 mM Na_2HPO_4, and 1.8 mM KH_2PO_4 (pH 7.4).

4. Extraction buffer: prepare according to Tables 1 and 2. 2xPre-TGEM buffer is autoclaved and stored at 4 °C, but extraction buffer is freshly prepared.

5. Glutathione Sepharose 4B agarose beads (GE Healthcare Life Sciences).

6. 10 mg/mL lysozyme.

7. Elution buffer: 10 mM GSH in 50 mM Tris–HCl (pH 8.0).

8. Amicon Ultra-4 centrifugal filter unit with Ultracel-30 membrane (Millipore).

9. Incubator shaker at 37 °C.

Table 1
2xPre-TGEM buffer

Stock	Stock concentration	Volume (mL)	Final concentration
Tris 8.0	1 M	8	40 mM
Glycerol		80	40 %
EDTA	0.5 M	0.8	2 mM
$MgCl_2$	1 M	2	10 mM
NP-40		0.4	0.2 %
H_2O		108.8	
		Total 200	

Table 2
Extraction buffer

Stock	Stock concentration	Volume	Final concentration
Pre-TGEM	2×	50 mL	
NP-40		0.9 mL	1 %
NaCl	5 M	3 mL	150 nM
DDT	1 M	100 μL	0.1 M
PMSF	0.2 M	500 μL	1 mM
Benzonase nuclease	0.3 M	100 μL	300 nM
H_2O		45.4 mL	
		Total 100 mL	

2.5 Kinase Assays

1. PKA and CK1 kinases are from New England Biolabs. Gprk2 kinase is purified by immunoprecipitation with the anti-Flag antibody using S2 cells transfected with Flag-Gprk2 or Flag-Gprk2 mutant. Recombinant human PKCζ is obtained from CD Biosciences.

2. Reaction buffer for in vitro kinase assays. For PKA and CK1 kinases: 50 mM Tris–HCl (pH 7.5), 10 mM $MgCl_2$, and 5 mM DTT. For Gprk2 kinase assay: 20 mM Tris–HCl (pH 8.0), 2 mM EDTA, 10 mM $MgCl_2$, and 1 mM DTT. For PKCζ kinase assay: 35 mM Tris–HCl (pH 7.5), 10 mM $MgCl_2$, 0.1 mM $CaCl_2$, 0.5 mM EGTA, and 2.5 μM ATP or 10 μCi of γ-^{32}P-ATP if needed.

3. Phosphatase inhibitor: 50 nM okadaic acid (OA, Calbiochem).

4. Kinase inhibitors. PKC inhibitor: PKCζ pseudosubstrate, myristoylated (Millipore). CK1 inhibitor: CK1-7 dihydrochloride (Sigma). PKA inhibitor: H-89 dihydrochloride hydrate (Sigma). *See* **Note 1** for working conditions.

5. Lysis buffer for cultured S2 cells: 50 mM Tris–HCl (pH 8.0), 100 mM NaCl, 1.5 mM EDTA, 10 % glycerol, 1 % NP-40, 10 mM NaF, 1 mM Na_3VO_4, and protease inhibitor tablet (Roche).

6. Protein A UltraLink Resin (Thermo).

7. Loading buffer (4×): 0.25 M Tris–HCl (pH 6.8), 8 % SDS, 20 % 2-mercaptoethanol, 40 % glycerol, and 0.008 % bromophenol blue.

8. Running buffer: 25 mM Tris base, 190 mM glycine, and 0.1 % SDS.

9. Transfer buffer: 25 mM Tris base, 190 mM glycine, and 20 % methanol.

10. Polyvinylidene difluoride (PVDF) membranes (Millipore).

11. TBST solution: 20 mM Tris–HCl (pH 7.6), 150 mM NaCl, and 0.1 % Tween-20.

12. Blocking buffer: 5 % nonfat dried milk or bovine serum albumin (BSA) in TBST.

13. Chemiluminescence detection reagents: Immobilon Western Chemiluminescent HRP substrate (Millipore), including luminol reagent and peroxide solution.

14. Blue X-Ray Film (Phenix Research).

15. Stripping buffer: 62.5 mM Tris–HCl (pH 6.8), 100 mM 2-mercaptoethanol, and 2 % SDS.

16. Antibodies: rabbit anti-SmoP (1:10), mouse anti-SmoN (Developmental Studies Hybridoma Bank, DSHB, 1:10), rabbit anti-GST (Santa Cruz, 1:10,000), mouse anti-Flag M2 (Sigma, 1:1,000), mouse anti-Myc, 9E10 (Santa Cruz, 1:50), rabbit anti-GFP (Millipore, 1:1,000; *see* **Note 2**), HRP-conjugated secondary antibodies (Jackson ImmunoResearch Laboratories).

2.6 Cell Surface Staining and Wing Disc Immunostaining

1. Coverslips and glass slides.

2. 4 % formaldehyde prepared freshly from a 37 % stock.

3. Blocking sera: normal donkey serum, normal goat serum (Jackson ImmunoResearch Laboratories).

4. PBST (PBS supplemented with 1 % Triton X100).

5. Antibodies: mouse anti-SmoN (DSHB, 1:10), epitope-tag antibodies to monitor the cotransfected proteins, mouse anti-Ptc (DSHB, 1:10), mouse anti-En (DSHB, 1:10), rabbit anti-β-Gal (Cappel, 1:1,500), and minimum cross-reactive (min X) secondary antibodies (Jackson ImmunoResearch Laboratories).

6. VECTASHIELD Mounting Medium H-1000 (Vector Laboratories, Inc).

7. Fingernail polish.

8. Olympus FluoView 1000 confocal microscope.

3 Methods

3.1 Luciferase Reporter Assay

1. Seed 2×10^5 S2 cells per well in a 48-well plate (in 250 μL Schneider's medium) and let cells attach and grow for 16 h in a humidified incubator at 25 °C.

2. Transfect the cells using Effectene transfection reagent according to the manufacturer's instructions. Briefly, prepare

the transfection mixture in an Eppendorf tube: 100 μL buffer EC, 5 ng Renilla, 150 ng Ub-Gal4, 50 ng tub-Ci (*see* **Note 3**), and 150 ng *ptc*-luciferase reporter constructs with or without pUAST-Smo constructs. The total amount of DNA transfected must be kept constant with the empty pUAST plasmid. Then add 4 μL enhancer and mix by vortexing. Incubate at room temperature for 5 min and add 6 μL Effectene transfection reagent to the mixture. Mix thoroughly and incubate for another 15 min. Add the mixture onto the cells and gently swirl the plate to distribute the transfection complex.

3. Incubate for 24 h, and then replace culture medium for fresh S2 medium (negative control) or S2 medium containing 60 % Hh-conditioned medium. Incubate for 24 h more.

4. 48 h post-transfection, lyse the cells for luciferase activity analysis using the Dual-Luciferase Reporter Assay System. Carefully remove the cell culture medium from the culture plate and rinse S2 cells with PBS. Then add 60 μL passive lysis buffer per well and gently shake the culture plate for 20 min at room temperature.

5. Measure the signal of the two luciferases using a GloMax Multi Detection System (Promega) in a 96-well plate, with 20 μL lysate in each well. Set the machine to dispense 100 μL Luciferase Assay Substrate to measure firefly luciferase activity, and then dispense 100 μL Stop & Glo reagent to measure Renilla luciferase activity. Renilla luciferase activity will be used to normalize the *ptc*-luciferase activity. Each sample should be tested at least in quadruplicate, and a representative assay should come from three independent experiments.

3.2 In Vitro Kinase Assay

3.2.1 Preparation of GST-Smo C-Tail Fusion

1. Inoculate single colony of glutathione S-transferase (GST)-Smo aa656-755 in *E. coli* BL21 (DE3) into LB medium with 100 μg/mL ampicillin. Incubate overnight at 37 °C in a shaker at 200 rpm.

2. Dilute the culture 1:100 into the fresh LB medium and grow until the culture OD600 value reaches 0.5–0.6. Add 0.1 mM IPTG and continue to incubate at 30 °C for 3–4 h (*see* **Note 4**).

3. Pellet the cells at $1,500 \times g$ at 4 °C, then resuspend cells in 15 mL extraction buffer containing 5 μg/mL lysozyme and incubate on ice for 30 min. All steps should be performed on ice or at 4 °C from here on.

4. Sonicate the suspension until it becomes clear, at 35 % output, alternating 8 s on/8 s off (16 min for a 15 mL sample). Increase length of time for larger samples.

5. Spin for 10 min at $13,500 \times g$ at 4 °C and transfer the supernatant to a tube containing 500 μL pre-equilibrated Glutathione

Sepharose 4B agarose beads per liter of bacterial culture. Rock at 4 °C overnight (*see* **Note 5**).

6. Centrifuge at $100 \times g$ at 4 °C for 5 min, discard the supernatant, and wash the beads three times with chilled extraction buffer.

7. After the last wash, remove the extraction buffer and add 500 µL of elution buffer. Rock at 4 °C for 30 min.

8. Centrifuge the sample at $100 \times g$ at 4 °C for 5 min. The supernatant is stored, and the elution is repeated for four to five times. These eluates can be pooled together and concentrated with centrifugal filter unit with Ultracel-30 membrane (Millipore).

3.2.2 In Vitro Phosphorylation Assays

Recombinants of purified PKA, CK1, Gprk2 (purified from S2 cells), and PKC kinases are used to phosphorylate GST-Smo fusion proteins in vitro, followed by analysis by Western blot or autoradiography (if γ-^{32}P-ATP is used in the phosphorylation reaction). When using radioisotopes, make sure to shield the samples with a Plexiglas screen and to survey the work area immediately afterward.

1. Incubate GST-Smo with individual kinases at 30 °C for 30 min in 50 µL of the kinase-specific reaction buffer. Stop the reaction by adding 4× SDS loading buffer to the sample.

2. Boil the sample at 100 °C for 5 min and centrifuge at $13,500 \times g$ in a microcentrifuge at 4 °C for 5 min. Samples can be stored at −20 °C and can be aliquoted to avoid repeat freeze/thaw cycles.

3. Load equal amounts of protein into the wells of an SDS-PAGE gel, along with molecular weight markers. Run the gel for 1–2 h at 90 V, and transfer the protein from the gel to the PVDF membrane at 180 mA for 3 h at 4 °C (*see* **Note 6**). The gel should be on the cathode side, and the membrane on the anode side of the transfer sandwich.

4. Block the membrane for 1 h using 5 % blocking solution, then incubate membrane with appropriate dilutions of anti-SmoP for 2 h at room temperature (*see* **Note 7**). Wash the membrane three times with TBST, 10 min each, followed by incubation with recommended dilution of labeled secondary antibody in 5 % blocking solution at room temperature for 1 h.

5. Wash the membrane three times with TBST, 10 min each, then mix equal volumes of luminol reagent and peroxide solution in a clean tube (*see* **Note 7**), and add it onto the membrane of the blot protein side. Incubate for 5 min and cover the membrane with a plastic wrap, drain the excess substrate and remove the air bubbles. Put an X-ray film onto it in the autoradiography cassette for 1 min (*see* **Note 8**). The anti-SmoP antibody can recognize the phosphorylated forms of Smo.

6. Submerge the membrane in stripping buffer and incubate at 50 °C for 30 min with occasional agitation to remove any bound antibodies, and then wash the membrane for 10 min in TBST three times at room temperature (*see* **Note 9**).

7. Re-block the membrane in 5 % nonfat dried milk in TBST for 1 h at room temperature and proceed with the standard Western blot protocol if wanted, such as to evaluate total GST-Smo, and proceed to expose the membrane to a new X-ray film for autoradiography if γ-^{32}P-ATP is used in the phosphorylation reaction. Expose at −80 °C for 24 h, and adjust further exposure based on the initial signal.

3.3 In Vivo Kinase Assay

1. Seed 1×10^6 S2 cells per well in a 6-well plate using 2 mL Schneider's medium per well and let cells grow for 16 h in a 25 °C incubator. Add 18 μg double-stranded RNA (dsRNA) to the medium to treat S2 cells if depletion of an endogenous protein is desired.

2. Transfect 350 ng attB-Myc-Smo and 150 ng Ub-Gal4 using Effectene transfection reagent like in Subheading 2.1, **step 2**. Of note, any pUAST-Smo construct can be used in this assay.

3. Incubate for 24 h, and then change cell culture medium into either control medium or medium containing 60 % Hh-conditioned medium for additional 24 h. Kinase inhibitor treatment for S2 cells is used if needed (*see* **Note 10**).

4. 48 h post-transfection, harvest S2 cells and add 450 μL lysis buffer. Then obtain cell lysate by centrifuging at $13,500 \times g$ for 10 min at 4 °C and separating the supernatant.

5. The lysate is subjected to immunoprecipitation (IP). Add primary antibody to the cell lysates and incubate at 4 °C for 2 h, and then add 30 μL protein A UltraLink Resin.

6. Wash the resin with wash buffer 5 min for three times on rocker at 4 °C. Then obtain 30 μL IP samples by centrifuging at $1, 200 \times g$ for 3 min at 4 °C, followed by adding 4× SDS loading buffer to the IP sample. 5 μL IP sample is loaded for each run.

7. Carry out Western blot with the method described in previous subsection. Hh consistently induces Smo phosphorylation, which can be readily detected by the anti-SmoP antibody. For better distinguishing phosphorylated forms from the unphosphorylated form of Smo, run 8 % separating gel until the 55 kDa marker is about to reach the bottom of the gel.

8. Electrophoretic mobility shift can also be used to detect the Smo phosphorylation. Different levels of Hh activity induce differential phosphorylation of Smo (*see* **Note 11** and Fig. 1). The OA treatment results in peak levels of phosphorylation in mobility shift of the transfected Myc-Smo.

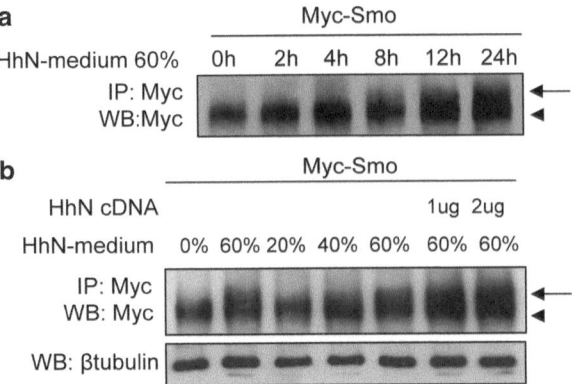

Fig. 1 Smo is differentially phosphorylated in response to Hh activity gradient. (**a**) S2 cells were transfected with Myc-Smo and treated with 60 % Hh-conditioned medium for different time periods. Cell extracts were immunoprecipitated with anti-Myc antibody followed by Western blot with the anti-Myc antibody. (**b**) S2 cells were transfected with Myc-Smo together with different amounts of Hh cDNA. Cell extracts were immunoprecipitated with anti-Myc antibody followed by Western blot with the anti-Myc antibody. Cell lysates were also subjected to Western blot with anti-tubulin, which serves as a loading control. *Arrows* indicate the hyperphosphorylated forms, and arrowheads indicate the unphosphorylated forms of Smo

3.4 Cell Surface Staining Assay

1. Sterilize the coverslips in 95 % ethanol and air dry, then put them in a sterile petri dish containing 50 μg/mL poly-L-lysine solution and incubate for 1 h at room temperature (15–25 °C). Wash with sterile water and allow to air dry.

2. Seed 1×10^6 S2 cells in 2 mL Schneider's medium per well in a 6-well plate containing one sterile coverslip in each well. Let cells grow for 16 h at 25 °C in a humidified incubator.

3. 200 ng pUAST-CFP-Smo and 150 ng Ub-Gal4 are transfected using Effectene transfection reagent. 24 h after transfection, change medium into either 60 % Hh-conditioned medium or control medium for 24 h.

4. Use a pipette to remove the cell culture medium from the plate well, and wash S2 cells with PBS 10 min for two times, at room temperature, followed by fixation with 4 % formaldehyde in PBS for 20 min. Then wash cells three times with PBS for 10 min each.

5. Add 1.5 % normal blocking serum in PBS for 30 min to block nonspecific staining. Blocking serum ideally should be derived from the same species in which the secondary antibody is raised.

6. Incubate the samples with the mouse anti-SmoN antibody (1:100) in 1 % normal blocking serum in PBS for 2 h before

cell permeabilization. The anti-SmoN antibody recognizes the extracellular domain of Smo [30].

7. Rinse S2 cells with 1 mL PBST 10 min for two times, at room temperature to allow cell permeabilization. Then incubate cells with Rhodamine-conjugated goat anti-mouse IgG (1:500) in 1 mL PBST for 1 h in dark chamber.

8. Wash three times with PBST for 10 min at room temperature in dark chamber.

9. Remove coverslips from the 6-well plate and mount on slides with the cell side toward the mounting medium, and seal the edges of the coverslip with fingernail polish.

10. Immunofluorescence is analyzed using a confocal microscope, such as Olympus FluoView 1000. Rhodamine signals represent the cell surface localized Smo, and CFP signals represent the total Smo that is expressed. Quantification of cell surface and total Smo levels can be calculated using MetaMorph software.

3.5 FRET Assay to Determine C-Terminal CFP/YFP Dimerization

Hh induces Smo phosphorylation and dimerization at its C-terminal tail [7, 8]. FRET assay can be used to test Smo C-tail dimerization.

1. Transfect S2 cells with Smo constructs tagged with CFP or YFP at the C-terminus and treat the cells with HhN-conditioned medium or control medium as previously described. For maximal Hh signal strength, a pUAST-Hh construct can be included in the transfection.

2. Wash the cells with PBS, fix them with 4 % formaldehyde for 20 min, and mount cells on slides in mounting medium.

3. Fluorescence signals can be acquired with the 60× objective on an Olympus confocal microscope. CFP is excited at 458 nm wavelength, and the emission is collected through a BA 480–495 nm filter. YFP is excited at 514 nm wavelength, and the emission can be collected through a BA 535–565 nm filter. The CFP signal is obtained once before (BP) and once after (AP) photobleaching YFP using the full power of the 515 nm laser line for 2–3 min at the top half of each cell, leaving the bottom half of the cell as an internal control. The intensity change of CFP can be analyzed using the Olympus FluoView software.

4. The efficiency of FRET will be calculated using the formula: $\text{FRET\%} = [(\text{CFP}_{AP} - \text{CFP}_{BP})/\text{CFP}_{AP}] \times 100$ %. Each data set should be based on 15 individual cells. In each cell, five regions of interest in the photobleached area should be selected for analysis [11].

3.6 Determination of Smo Mutants Activity in the Wing Disc of Transgenic Flies

To generate attB-UAST-Myc-Smo, the attB sequence [26] was first inserted into the Hind III site upstream of the UAS-binding sites in the pUAST vector (Fig. 2). Then, the Myc-Smo sequence was inserted into the multiple cloning sites (MCS). Smo mutants are generated by site-directed mutagenesis and subcloned into the attB-UAST vector. All constructs should be verified by sequencing and all plasmid DNA purified by maxiprep.

3.6.1 Generation of attB-Containing Constructs

3.6.2 Transgenic Flies Generation

Transgenic fly lines are generated with the ΦC31-based integration system to determine the activity of Smo variants in the wing disc. The *vas-phi-zh2A-VK5* flies are used to insert attB-UAST-Myc-Smo transgenes at the VK5 attP locus, which ensures the same level of expression without positional effects [29]. Smo mutant flies are generated by microinjection according to standard procedures. *VK5-tub-Flag-Smo*[WT], encoding wild-type Smo under the transgene *tubulin-α* promoter, was used to express Smo at low level at VK5 locus as positive control for this in vivo assay [9].

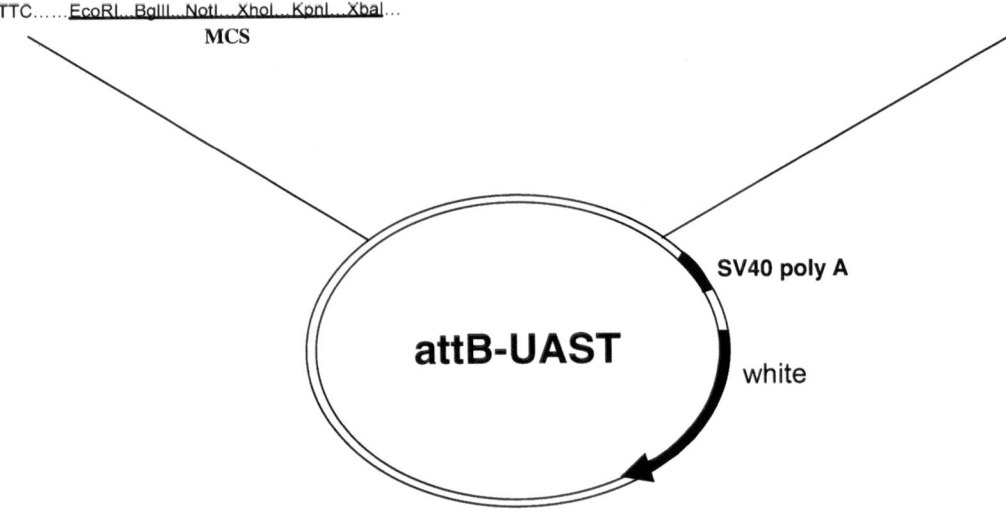

Fig. 2 attB-UAST vector. The region of multiple cloning sites (MCS) and special elements are *underlined*. The attB sequence is inserted into the Hind III site in upstream of UAS-binding sites

To evaluate the ability of *VK5-tub-Smo^{WT}* to rescue *smo* null phenotype in wing discs, larvae from the genotypes *y hsp-flp; smo³ FRT 39E/hsp-CD2, y+FRT39E; tub-smo^{WT}/dpp-lacZ* were subjected to heat shock to induce *smo* mutant clones. Late third instar wing discs were immunostained as detailed in the following section with anti-CD2 and anti-β-gal antibodies to visualize the expression of CD2 and *dpp-lacZ*. *smo* mutant clones were recognized by the lack of CD2 expression. *VK5-tub-Smo^{WT}* rescued *dpp-lacZ* expression in *smo* mutant clones, which validates the assay to compare the ability of the attB-UAST-Myc-Smo transgenes that one prefers to analyze.

3.6.3 Drosophila Wing Disc Immunostaining

1. Dissect the wing discs from third instar larvae with specific genotypes in PBS (for more detail, *see* Chapters 4 and 5).

2. Fix the wing discs with 4 % formaldehyde in PBS for 20 min.

3. Permeabilize with PBST and incubate the discs with the indicated primary antibodies for 2 h and the corresponding secondary antibodies for 1 h sequentially.

4. Perform three 20 min washes with PBST.

5. Mount the samples in a liquid media on a microscope slide.

4   Notes

1. The kinase inhibitors can be used to treat the S2 cells in cell medium; the effective conditions are in Table 3. For in vitro kinase assay, they can be added directly to the reaction buffer at proper concentration if needed.

2. pUAST-GFP construct can be cotransfected into S2 cells and used as a transfection control using the GFP antibody.

3. Because S2 cells do not express Ci [31], tub-Ci needs to be cotransfected to evaluate the activity of Smo mutants on *ptc-luciferase* expression, which is driven by Ci binding to the *ptc* promoter region [11]. Renilla is also transfected, and its luciferase activity is used to normalize the *ptc*-luciferase activity to account for variations in cell transfection efficiency.

Table 3
Kinase inhibitors

Inhibitor	Target kinase	Final concentration (μM)	Adding time before harvesting (h)
PKCζ pseudosubstrate, myristoylated	PKC	10	4
CK1-7 dihydrochloride	CK1	10	4
H-89 dihydrochloride hydrate	PKA	10	4

Table 4
Gel recommendations

Protein size (kDa)	Gel percentage (%)
4–40	20
12–60	15
20–140	12.5
30–200	10
40–300	8

4. Inducing expression with 0.1 mM IPTG at 30 °C for 3 h usually yields considerable amount of soluble protein or otherwise may form inclusion body aggregates. Lower temperature should be considered if the proteins are not soluble.

5. Equilibrate the glutathione agarose beads with extraction buffer before loading supernatant onto them.

6. The time and voltage may require some optimization, and gel percentage will depend on the size of the Smo fragment of interest (for a guide, *see* Table 4).

7. Incubating overnight at 4 °C is also advisable. Make sure that the solution moves freely across the entire surface of the membrane.

8. Approximately 100 μL of working HRP substrate is required per cm² membrane area.

9. For getting a clear signal, the appropriate duration time can be varied; the initial exposure time is 30 s. The chemiluminescent signal on the membrane will last 30 min, and fresh HRP substrate can be added to the same blot for consecutive exposures.

10. The incubation can be performed at 70 °C or the incubation time increased to more stringent conditions. To verify the removal of the previous antibody, the membrane could be detected with the secondary antibody.

11. For showing the mobility shift of Smo, low voltage (about 90 V) and extended running time are recommended. The loading sample could be heated at 55 °C for 5 min to reduce Smo aggregation.

Acknowledgment

We thank the members in Dr. Jia's Lab for their efforts to this protocol. This work was supported by grant from the National Institutes of Health (GM079684) to J. Jia.

References

1. Nusslein-Volhard C, Wieschaus E (1980) Mutations affecting segment number and polarity in Drosophila. Nature 287:795–801

2. Ingham PW, Nakano Y, Seger C (2011) Mechanisms and functions of Hedgehog signalling across the metazoa. Nat Rev Genet 12: 393–406

3. Jiang J, Hui CC (2008) Hedgehog signaling in development and cancer. Dev Cell 15: 801–812

4. Ingham PW, McMahon AP (2001) Hedgehog signaling in animal development: paradigms and principles. Genes Dev 15:3059–3087

5. Zhang C, Williams EH, Guo Y, Lum L, Beachy PA (2004) Extensive phosphorylation of Smoothened in Hedgehog pathway activation. Proc Natl Acad Sci U S A 101:17900–17907

6. Apionishev S, Katanayeva NM, Marks SA, Kalderon D, Tomlinson A (2005) Drosophila Smoothened phosphorylation sites essential for Hedgehog signal transduction. Nat Cell Biol 7:86–92

7. Jia J, Tong C, Wang B, Luo L, Jiang J (2004) Hedgehog signalling activity of Smoothened requires phosphorylation by protein kinase A and casein kinase I. Nature 432:1045–1050

8. Zhao Y, Tong C, Jiang J (2007) Hedgehog regulates smoothened activity by inducing a conformational switch. Nature 450:252–258

9. Jia H, Liu Y, Xia R, Tong C, Yue T, Jiang J, Jia J (2010) Casein kinase 2 promotes Hedgehog signaling by regulating both smoothened and Cubitus interruptus. J Biol Chem 285: 37218–37226

10. Chen Y, Li S, Tong C, Zhao Y, Wang B et al (2010) G protein-coupled receptor kinase 2 promotes high-level Hedgehog signaling by regulating the active state of Smo through kinase-dependent and kinase-independent mechanisms in Drosophila. Genes Dev 24: 2054–2067

11. Fan J, Liu Y, Jia J (2012) Hh-induced Smoothened conformational switch is mediated by differential phosphorylation at its C-terminal tail in a dose- and position-dependent manner. Dev Biol 366:172–184

12. Jiang K, Liu Y, Fan J, Epperly G, Gao T, Jiang J, Jia J (2014) Hedgehog-regulated atypical PKC promotes phosphorylation and activation of Smoothened and Cubitus interruptus in Drosophila. Proc Natl Acad Sci U S A 111:E4842–E4850

13. Rohatgi R, Milenkovic L, Scott MP (2007) Patched1 regulates hedgehog signaling at the primary cilium. Science 317:372–376

14. Rohatgi R, Milenkovic L, Corcoran RB, Scott MP (2009) Hedgehog signal transduction by Smoothened: pharmacologic evidence for a 2-step activation process. Proc Natl Acad Sci U S A 106:3196–3201

15. Wang Y, Zhou Z, Walsh CT, McMahon AP (2009) Selective translocation of intracellular Smoothened to the primary cilium in response to Hedgehog pathway modulation. Proc Natl Acad Sci U S A 106:2623–2628

16. Corbit KC, Aanstad P, Singla V, Norman AR, Stainier DY, Reiter JF (2005) Vertebrate Smoothened functions at the primary cilium. Nature 437:1018–1021

17. Chen Y, Sasai N, Ma G, Yue T, Jia J, Briscoe J, Jiang J (2011) Sonic Hedgehog dependent phosphorylation by CK1alpha and GRK2 is required for ciliary accumulation and activation of smoothened. PLoS Biol 9:e1001083

18. Wilson CW, Chuang PT (2010) Mechanism and evolution of cytosolic Hedgehog signal transduction. Development 137:2079–2094

19. Basler K, Struhl G (1994) Compartment boundaries and the control of Drosophila limb pattern by hedgehog protein. Nature 368: 208–214

20. Tabata T, Kornberg TB (1994) Hedgehog is a signaling protein with a key role in patterning Drosophila imaginal discs. Cell 76:89–102

21. Lecuit T, Brook WJ, Ng M, Callega M, Sun H, Cohen SM (1996) Two distinct mechanisms for long-range patterning by Decapentaplegic in the Drosophila wing. Nature 381:387–393

22. Nellen D, Burke R, Struhl G, Basler K (1996) Direct and Long-Range Action of a DPP Morphogen Gradient. Cell 85:357–368

23. Campbell G, Tomlinson A (1999) Transducing the Dpp morphogen gradient in the wing of Drosophila: regulation of Dpp targets by brinker. Cell 96:553–562

24. Jia J, Jiang J (2006) Decoding the Hedgehog signal in animal development. Cell Mol Life Sci 63:1249–1265

25. Hooper JE, Scott MP (2005) Communicating with Hedgehogs. Nat Rev Mol Cell Biol 6:306–317

26. Bateman JR, Lee AM, Wu CT (2006) Site-specific transformation of Drosophila via phiC31 integrase-mediated cassette exchange. Genetics 173:769–777

27. Venken KJ, He Y, Hoskins RA, Bellen HJ (2006) P[acman]: a BAC transgenic platform for targeted insertion of large DNA fragments in D. melanogaster. Science 314:1747–1751

28. Bischof J, Maeda RK, Hediger M, Karch F, Basler K (2007) An optimized transgenesis system for Drosophila using germ-line-specific phiC31 integrases. Proc Natl Acad Sci U S A 104:3312–3317

29. Jia H, Liu Y, Yan W, Jia J (2009) PP4 and PP2A regulate Hedgehog signaling by controlling Smo and Ci phosphorylation. Development 136:307–316

30. Lum L, Zhang C, Oh S, Mann RK, von Kessler DP, Taipale J et al (2003) Hedgehog signal transduction via Smoothened association with a cytoplasmic complex scaffolded by the atypical kinesin, Costal-2. Mol Cell 12:1261–1274

31. Denef N, Neubuser D, Perez L, Cohen SM (2000) Hedgehog induces opposite changes in turnover and subcellular localization of patched and smoothened. Cell 102:521–531

Chapter 6

Investigation of Protein–Protein Interactions and Conformational Changes in Hedgehog Signaling Pathway by FRET

Lin Fu, Xiangdong Lv, Yue Xiong, and Yun Zhao

Abstract

Protein–protein interactions and signal-induced protein conformational changes are fundamental molecular events that are considered as essential in modern life sciences. Among various techniques developed to study such phenomena, fluorescence resonance energy transfer (FRET) is a widely used method with many advantages in detecting these molecular events. Here, we describe the application of FRET in the mechanistic investigation of cell signal transduction, taking the example of the Hh signaling pathway, which plays a critical role in embryonic development and tissue homeostasis. A number of general guidelines as well as some key notes have been summarized as a protocol for reader's reference.

Key words FRET, Acceptor bleach, Hh signaling pathway, Conformational changes, Protein–protein interactions

1 Introduction

FRET is a well-established technique developed for study of protein–protein interactions and signal-induced protein conformational changes. The principle of FRET is to trace the variation of distance between atoms or molecules through measuring the energy transfer from an excited donor to an acceptor [1, 2]. If the distance between two molecules is less than 10 nm, there will be an energy transfer from the donor to the acceptor when the donor is excited. There are several different methods used to detect FRET events, including "conventional filter FRET" and "acceptor bleach." In "conventional filter FRET," application of filter/emission band configurations for donor, acceptor, and FRET (donor excitation and acceptor emission) is used to acquire single or time series images. In this case of FRET, the donor signal decreases, and acceptor and FRET signal increases. "Acceptor bleach" is by applying

Natalia A. Riobo (ed.), *Hedgehog Signaling Protocols*, Methods in Molecular Biology, vol. 1322,
DOI 10.1007/978-1-4939-2772-2_6, © Springer Science+Business Media New York 2015

donor–acceptor configurations to acquire single or time series images. After some control images, the acceptor is bleached. In such case of FRET, donor signal increases after acceptor bleach. In this protocol, we focus on the method of acceptor bleach. By measuring the fluorescence intensity changes of the donor before and after photobleaching and using the formula $\text{FRET }\% = [(\text{donor}^{\text{after}} - \text{donor}^{\text{before}})/\text{donor}^{\text{after}}] \times 100$, we can detect the percentage change of donor fluorescence intensity, which indicates whether the two molecules are physically adjacent to each other, and even evaluate the approximate distance between them (Fig. 1). In theory, for one protein to interact with another, they have to be in close proximity. Therefore, FRET is frequently used to investigate protein–protein interactions and protein conformational changes, especially during the mechanistic dissection of signal transduction (Fig. 2).

The advantages of FRET relative to other techniques include at least two aspects. First, when compared with other methods to detect protein–protein interactions such as immunoprecipitation and GST pull-down, FRET is more sensitive to small distance and can be easily applied in vivo, i.e., in cells or tissues. Second, when compared with circular dichroism and 3D structure determination that require purified protein to detect protein conformational changes, FRET is easier to carry out since transfection system works well. On the other hand, there are limitations of FRET. For example, the diameter of the target molecules used for measuring

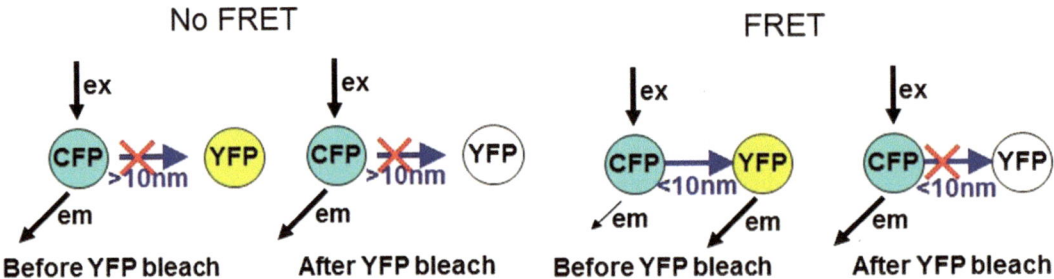

Fig. 1 The principle of fluorescence resonance energy transfer (FRET). FRET is the nonradioactive transfer of photon energy from an excited fluorophore (the donor) to another fluorophore (the acceptor) when both are located within close proximity (1–10 nm). And it is used to resolve questions such as molecular interactions between two proteins or domains, which are beyond the optical limit of a light microscope. An excited fluorophore (CFP, the donor) transfers its excited state energy to a light-absorbing molecule (YFP, the acceptor). If CFP and YFP are not located within close proximity (more than 10 nm), there is no energy transfer. In this case, if YFP is photobleached, intensity of CFP will not change. If the two fluorophores are close to each other (<10 nm), the energy will be transferred from CFP to YFP. In this case, when YFP is photobleached, the energy transfer will be inhibited, and then the intensity of CFP will increase after YFP bleached. By using the formula $\text{FRET }\% = [(\text{donor}^{\text{after}} - \text{donor}^{\text{before}})/\text{donor}^{\text{after}}] \times 100$, one can know if the two molecules are located with close proximity and even to evaluate the distance

a
Intramolecular FRET (domain A and B)

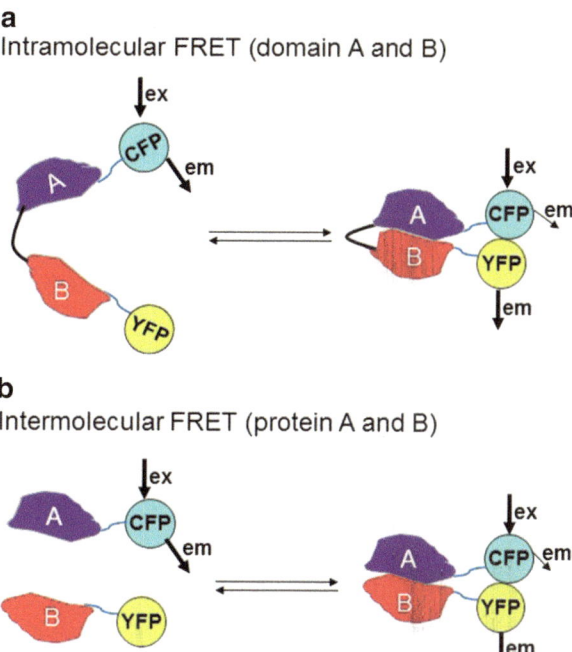

b
Intermolecular FRET (protein A and B)

Fig. 2 Applications of FRET. FRET can be used to investigate protein conformational change (intramolecular FRET) and protein–protein interaction (intermolecular FRET). (**a**) To detect the protein conformational change, the donor (CFP) and acceptor (YFP) could be fused to domains A and B of the same protein, respectively. If the protein keeps an "open" conformation, the donor (CFP) will not be located close proximity from the acceptor (YFP). Energy transfer will not happen and FRET could not be detected. If there is conformational change, the domain A of host protein will become close to B (less than 10 nm), and energy transfer will happen. In this case, FRET could be detected by measuring the intramolecular fluorescence intensity changes. (**b**) To detect protein–protein interactions, the donor (CFP) and acceptor (YFP) could be fused to N- or C-termini of protein A and B, respectively. By detecting the FRET of intermolecular reading, we can know whether the two proteins interact or not under signal stimulation

FRET should not be bigger than 2–10 nm. Meanwhile, as the donor emission spectrum has to overlap the excitation spectrum of acceptor, only a few pairs of fluorophores are suitable for FRET experiments, such as CFP/YFP, BFP/GFP, GFP/rhodamine, and FITC/Cy3.

It has been well established that the evolutionarily conserved Hedgehog (Hh) signaling pathway plays a fundamental role in embryonic development and tissue homeostasis and that dysregulation of Hh pathway is closely associated with various diseases including birth defect and cancer [3–6]. As a powerful method to investigate protein–protein interactions and protein conformational changes, it is not surprising that FRET has been widely used for studying the mechanism of Hh signaling since a decade ago. A number of scientific breakthroughs on the Hh signaling transduction have been achieved by FRET assay combined with certain functional studies. For example, Hh proteins form oligomers on the Hh-producing cell's surface, which aid in the interaction with heparan sulfate proteoglycans (HSPGs) and its long-range transport [7]. *Drosophila* Patched (Ptc) forms a homo-trimer, which is necessary for its internalization and turnover [8]. In *Drosophila*, the conformational changes of Smoothened (Smo) C-terminus are essential for the transduction of the Hh signal. In the absence of Hh, Smo stays in a "close" conformation recognized as an inactive form, while in the presence of Hh, Smo C-terminus is phosphorylated and turns into an "open" conformation recognized as an active form, which further leads to activation of the Hh signaling pathway [9]. A similar mechanism is also observed in mammalian Smo [10]. Meanwhile, oligomerization/higher-order clustering of Smo has also been identified by FRET in lipid rafts responding to high level of Hh signal [11]. Moreover, Smo phosphorylation and conformational change further induce dimerization of Fused, which then transduces the signal by phosphorylating Costal2 (Cos2) and Suppressor of Fused (Sufu) [12–14]. The conformational change of Sufu between "close" and "open" states is also important for the Hh signaling. Specifically, Sufu binds to glioma-associated oncogene homologue (Gli) in a "close" conformation, which decreases Hh signaling, while its "open" conformation promotes Hh signaling by releasing Gli [15].

Despite of these important FRET-related discoveries on Hh signaling, there are still numerous open questions to address. For example, the mechanism through which Ptch transduces Hh signaling to Smo and then Smo responds to the Hh gradient is still unclear. Apparently, FRET in combination with other techniques may be used in this line of research to screen the missing link between Ptch and Smo. According to the structural analysis, Smo response to Hh signaling may be related to the conformational changes of its extracellular linker domain [16, 17], which can be tested by single-molecule FRET. Meanwhile, drug discovery targeting Hh pathway is of intense interest for pharmaceutical industry. Although the first drug targeting Smo, vismodegib, has been approved by the US Food and Drug Administration in 2012, the treatment-driven evolution has resulted in drug resistance [18]. In this regard, FRET-based high-throughput screening will be an attractive choice to find out suitable drug candidates as well as to improve the current medical treatment.

Here, we describe the standard lab-use protocol for FRET assay in fly and mammalian systems for both cell-based and tissue-based experiments.

2 Materials

2.1 Equipment

1. 25 °C, 37 °C incubators (Thermo).
2. 24-well plate and 10 cm culture dishes (Greiner Bio-One).
3. Glass slides (diameter 14.5 mm) (*see* **Note 1**).
4. Rotator.
5. Vacuum pump.
6. Dissection microscope (Leica).
7. Dissection forceps.
8. Leica SP5 confocal microscope.
9. Filter for the donor and acceptor.
10. 1.5 ml and 15 ml tubes (Axygen).

2.2 Cell Lines and Flies

1. S2 cells, commercially available from various sources.
2. Hh-N-producing cells [19].
3. NIH/3T3 cells (ATCC).
4. Indicated fly strains (*Drosophila melanogaster*).

2.3 Cell Culture Media

1. S2 cell culture medium: Schneider's Drosophila Medium (Invitrogen) containing 10 % fetal bovine serum (FBS) and 10 mL/L penicillin–streptomycin (5,000 U penicillin and 5000 µg streptomycin/mL).
2. NIH/3T3 cell culture medium: Dulbecco's Modified Eagle's Medium containing 10 % newborn calf serum (NCS) and 10 mL/L penicillin–streptomycin (5,000 U penicillin and 5,000 µg streptomycin/mL).

2.4 Reagents

1. FuGENE 6 (Promega).
2. Lipofectamine™ 2000 Transfection Reagent (Invitrogen).
3. Aqua-Poly/Mount Coverslipping Medium (Polyscience).
4. Phosphate buffered saline (PBS): 4.3 mM Na_2HPO_4, 137 mM NaCl, 2.7 mM KCl, 1.4 mM KH_2PO_4.
5. 0.7 M $CuSO_4$.
6. 37 % formaldehyde.
7. 80 % glycerin in water.
8. Nail polish.
9. Oil (Leica).

2.5 Plasmids Used for the Expression of Specific Proteins in S2 and NIH/3T3 Cells

1. For S2 cells: pUAST vectors (pUAST-Smo-CFP^C and pUAST-Smo-YFP^C).
2. For NIH/3T3: pcDNA vectors.

3 Methods

The method is divided into two parts: sample preparation and image scanning to get the FRET readings. We use YFP/CFP as an example to illustrate the method.

3.1 Sample Preparation

3.1.1 Drosophila S2 Cells

1. First, prepare Hh-conditioned medium to later stimulate the cells. Plate 4×10^7 Hh-producing S2 cells (HhN-stable cell line) in 10 ml medium, and 24 h later, add 10 µl 0.7 M $CuSO_4$ to induce Hh expression. Use cells without $CuSO_4$ induction to generate "control medium." The cells are cultured in 25 °C incubator for 24 h. Centrifuge the cells in a 15 ml tube for 5 min at $1,000 \times g$ and transfer the supernatant to a new tube. The conditioned medium can be stored at 4 °C for at most 1 week.

2. Second, prepare the cells for FRET analysis. Plate S2 cells in a 10 cm dish in 10 ml complete medium (10^6 cells/ml) and culture for 24 h in humidified incubator at 25 °C (*see* **Note 2**).

3. When cells reach 60–70 % confluency, transfect plasmids with FuGENE 6 following the manufacturer's procedure (*see* **Note 3**).

4. Treat the cells with Hh-conditioned medium according to the experiment design. Briefly, 24 h after transfection, transfer the cells that grow loosely attached to the dish into a 15 ml tube and centrifuge for 5 min at $1,000 \times g$. For the Hh treated groups, use 6 ml Hh conditional medium and 4 ml fresh S2 medium to resuspend the cells. For the control groups, use 6 ml control medium and 4 ml fresh S2 medium instead. The cells are plated in 10 cm dishes and cultured in a 25 °C incubator for 24 h or the desired time.

5. Use a pipette to transfer the cells from the dish to a 15 ml tube. Centrifuge the cells for 5 min at $1,000 \times g$ 4 °C. Discard the supernatant and resuspend the cells using 1 ml ice-cold PBS.

6. Then transfer the cells to a 1.5 ml tube and centrifuge for 5 min at $1,000 \times g$. To wash the cells, resuspend the cells in 1 ml ice-cold PBS and rotate for 5 min, and then centrifuge for 5 min and discard the supernatant. Repeat the wash step twice to ensure the wash is sufficient.

7. Fix the cells with 4 % formaldehyde in 1 ml PBS with rotation for 10 min. Then wash three times with PBS.

8. After the last centrifugation, discard the supernatant and resuspend in 20 µl PBS. Add another 20 µl 80 % glycerin and mix slightly. Put a drop of the mixture on a glass slide and add a coverslip on it (*see* **Note 4**). Use nail polish to seal the slides. Let it dry at room temperature and store at 4 °C.

3.1.2 Drosophila Wing Imaginal Discs	1. Set up fly crosses by using a driver to express the target gene in the specific location.

1. Set up fly crosses by using a driver to express the target gene in the specific location.

2. After about 7 days, dissect the last third instar larvae in 1 ml PBS.

3. Fix the larvae with 4 % formaldehyde in 1 ml PBS and rotate for 30 min.

4. Discard the PBS and formaldehyde mixture, and then wash with 1 ml PBS for 15 min, repeating three times.

5. Put a drop of PBS on a slide. Dissect the wing imaginal discs and use a pipette to transfer the discs into this drop.

6. Add 10 μl 80 % glycerin onto the drop and add a cover slide on it (*see* **Note 4**). Use nail polish to seal the slides. Let it dry at room temperature and store at 4 °C.

3.1.3 Mammalian Cells

1. Glass slides that fit into 24-well plates are prepared beforehand and stored in 75 % ethanol. Sterilize by flaming and then put them into the 24-well plate.

2. Plate 30,000 NIH/3T3 cells onto the glass slides, which are situated in the 24-well plate in 500 μl complete medium and culture for 24 h in humidified incubator at 37 °C.

3. When cells reach 60–70 % confluency, transfect the plasmids of interest using Lipofectamine 2000, following the manufacturer's instructions. Then culture for 24 h in 37 °C incubator.

4. Change the medium to DMEM containing 0.5 % NCS when the cells reach very high density to induce formation of primary cilia. And add drugs to the cells according to the experiment design and incubate for the desired amount of time.

5. Discard the medium and wash with 1 ml PBS per well. Rotate the plate 5 min in a rotator. Repeat this step twice more.

6. Fix the cells with 4 % formaldehyde in PBS for 10 min on the rotator (*see* **Note 5**).

7. Wash three times with 1 ml PBS per well for 5 min. Rotate slightly to improve washing.

8. Prepare a glass slide with two drops of Aqua-Poly/Mount on it. Take out the glass insert from the 24-well plate, which is covered with cells on its surface, and put it upside down on the drops to ensure the cells contact with the Aqua-Poly/Mount. Let it dry at room temperature and store at 4 °C.

3.2 FRET Detection

1. Use the Leica confocal SP5 microscope to detect the signal. Choose the model "FRET AB wizard."

2. Adjust the intensity of fluorescence and shoot the donor and acceptor images. The CFP signal is obtained by excitation at 458 nm and emission range from 416 to 492 nm. For imaging YFP fluorescence, the excitation wavelength is 514 nm, and the emission range is from 525 nm to 600 nm.

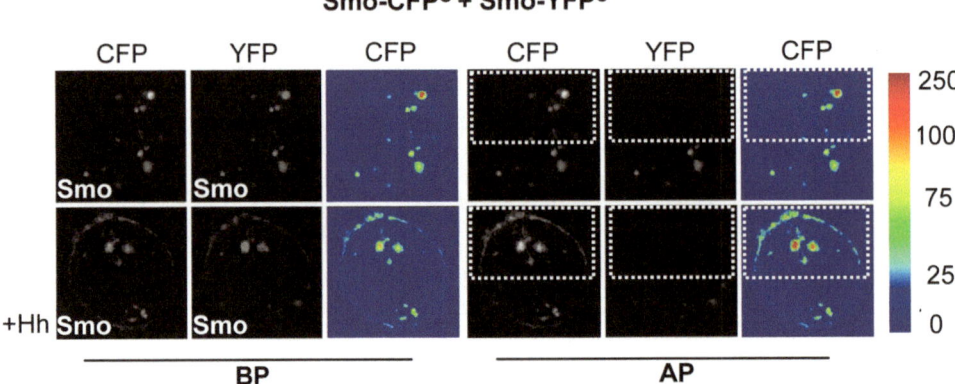

Fig. 3 FRET assay of S2 cells expressing Smo constructs (reproduced from Zhao et al. Nature, 2007). To detect FRET of Smo C-terminus interaction, S2 cells were transfected with Smo-CFP[C] and Smo-YFP[C], treated with or without Hh. Both in the absence and presence of Hh, Smo-CFP[C] and Smo-YFP[C] colocalized. CFP and YFP signals were acquired before (BP) and after (AP) photobleaching YFP at the top half of each cell. The unbleached area serves as internal control. Since the CFP was not bleached in the bottom half of cells, intensity of CFP should be constant before and after photobleaching YFP on the top half of cells. However, since S2 cells are semifloating cells, they might change position or angle on the slide after YFP bleaching. In this case, this will dramatically affect FRET result. To avoid this possibility, intensity of CFP at the bottom half of cell will be collected. If the intensity of CFP at the bottom half of cell did not change, FRET between Smo-CFP[C] and Smo-YFP[C] will be measured. If the intensity of CFP at the bottom half of cell changed, FRET between Smo-CFP[C] and Smo-YFP[C] should not be analyzed further. The pseudocolored bar with a color gradient from *blue* to *red* indicates CFP intensity from 0 to 250

3. Choose the bottom half as internal control (*see* **Note 6**), and bleach the top half (Fig. 3) [9]. To measure FRET, a few regions of interest are chosen to get an average reading of all cells. Continuous bleaching is achieved by using full power of the 514 nm wavelength until the acceptor fluorescence intensity is below background levels (*see* **Note 7**).

4. Get the result using LAS AF Lite software. By using the equation FRET $\% = [(\text{donor}^{\text{after}} - \text{donor}^{\text{before}})/\text{donor}^{\text{after}}] \times 100$, we can evaluate the distance between the two molecules. The final result should be the average of the readings from about 15–20 individual cells.

4 Notes

1. The size of glass is a little smaller than that of the 24 wells, which makes it easier to put in and take out. For cells adhere to the glass firmly, 0.01 % polylysine may need to be used to treat the glasses before use.

2. S2 cells is suspended cell. To maintain these cells, just use fresh S2 medium to dilute them at a ratio of 1:5 every 3 days.

3. For the S2 cell transfection, Effectene (Qiagen) and calcium phosphate-mediated transfection are also good choices.

4. As bubbles will affect image quality under the confocal scan, it is necessary to avoid their formation. To avoid bubbles, you might first clamp one side of the cover slide by nipper, let the other side contact with the drop of sample on the slide, and then put the cover slide flat on the sample gently. If there are still bubbles, you should put a few drops of PBS on one side of the cover slide with a pipette, and use absorbent paper to draw the PBS on the other side; repeat the process just until the bubbles are disappeared.

5. The cell should be fixed because the bleaching takes about 1–3 min; movement will dramatically affect the result and then reduces the sensibility of FRET. For the research about living cell FRET, you may choose to use fluorescence lifetime imaging (FLIM).

6. Because S2 cells and *Drosophila* discs are suspended in glycerin, they can drift during image acquisition, especially during bleaching, so it is necessary to ensure that the internal control stays the same. For the mammalian cells, this step is not necessary as they are firmly adhered to the slide if the cells are in good conditions.

7. In order to get an accurate result, sufficient bleaching for the acceptor is needed. Choosing an appropriate bleaching intensity and time is important. The bleaching time between the fly tissues and cells is different, but in general by using a high-density bleaching, the acceptor should be as weak as the background.

Acknowledgment

This work was supported by grants from the National Basic Research Program of China (973 Program: 2011CB943902), the "Strategic Priority Research Program" of the Chinese Academy of Sciences (XDA01010405).

References

1. Förster T (1948) Intermolecular energy migration and fluorescence. Ann Phys 2:55–75

2. Stryer L (1978) Fluorescence energy transfer as a spectroscopic ruler. Annu Rev Biochem 47:819–846

3. Ingham PW, McMahon AP (2001) Hedgehog signaling in animal development: paradigms and principles. Genes Dev 15:3059–3087

4. Rubin LL, De Sauvage FJ (2006) Targeting the Hedgehog pathway in cancer. Nat Rev Drug Discov 5:1026–1033

5. Jiang J, Hui CC (2008) Hedgehog signaling in development and cancer. Dev Cell 15:801–812

6. Hui CC, Angers S (2011) Gli proteins in development and disease. Annu Rev Cell Dev Biol 27:513–537

7. Vyas N, Goswami D, Manonmani A, Sharma P, Ranganath HA, VijayRaghavan K et al (2008) Nanoscale organization of hedgehog is essential for long-range signaling. Cell 133:1214–1227

8. Lu X, Liu S, Kornberg TB (2006) The C-terminal tail of the Hedgehog receptor Patched regulates both localization and turnover. Genes Dev 20:2539–2551

9. Zhao Y, Tong C, Jiang J (2007) Hedgehog regulates smoothened activity by inducing a conformational switch. Nature 450:252–258

10. Chen Y, Sasai N, Ma G, Yue T, Jia J, Briscoe J, Jiang J (2011) Sonic Hedgehog dependent phosphorylation by CK1alpha and GRK2 is required for ciliary accumulation and activation of smoothened. PLoS Biol 9:e1001083

11. Shi D, Lv X, Zhang Z, Yang X, Zhou Z, Zhang L, Zhao Y (2013) Smoothened oligomerization/higher order clustering in lipid rafts is essential for high Hedgehog activity transduction. J Biol Chem 288:12605–12614

12. Shi Q, Li S, Jia J, Jiang J (2011) The Hedgehog-induced Smoothened conformational switch assembles a signaling complex that activates Fused by promoting its dimerization and phosphorylation. Development 138:4219–4231

13. Tang JY, Xiao TZ, Oda Y, Chang KS, Shpall E, Wu A et al (2011) Vitamin D3 inhibits hedgehog signaling and proliferation in murine Basal cell carcinomas. Cancer Prev Res (Phila) 4:744–751

14. Zhang Y, Mao F, Lu Y, Wu W, Zhang L, Zhao Y (2011) Transduction of the Hedgehog signal through the dimerization of Fused and the nuclear translocation of Cubitus interruptus. Cell Res 21:1436–1451

15. Zhang Y, Fu L, Qi X, Zhang Z, Xia Y, Jia J et al (2013) Structural insight into the mutual recognition and regulation between Suppressor of Fused and Gli/Ci. Nat Commun 4:2608

16. Wang C, Wu H, Katritch V, Han GW, Huang XP, Liu W et al (2013) Structure of the human smoothened receptor bound to an antitumour agent. Nature 497:338–343

17. Rana R, Carroll CE, Lee HJ, Bao J, Marada S, Grace CR et al (2013) Structural insights into the role of the Smoothened cysteine-rich domain in Hedgehog signalling. Nat Commun 4:2965

18. Lv X, Fu L, Zhao Y (2013) aPKC iota/lambda: a potential target for the therapy of Hh-dependent and Smo-inhibitor-resistant advanced BCC. Acta Biochim Biophys Sin 45:610–611

19. Lum L, Zhang C, Oh S, Mann RK, von Kessler DP, Taipale J et al (2003) Hedgehog signal transduction via Smoothened association with a cytoplasmic complex scaffolded by the atypical kinesin, Costal-2. Mol Cell 12:1261–1274

Chapter 7

Luciferase Reporter Assays to Study Transcriptional Activity of Hedgehog Signaling in Normal and Cancer Cells

Silvia Pandolfi and Barbara Stecca

Abstract

The measurement of the transcriptional activity of the HH signaling pathway is widely used as an indication of pathway activation. Luciferase reporter assays are powerful tools to measure the specific ability of a transcription factor to bind to its consensus sequence and to activate transcription of target genes. Here, we describe a protocol to measure the transcriptional activity of the HH pathway in normal and cancer cells. This technique allows studying the activity of GLI transcription factors and their modulation by drugs and/or other factors.

Key words Hedgehog signaling, GLI, Transcription factor, Binding site, Luciferase, Transfection, Reporter assay

1 Introduction

Canonical HH pathway activation is initiated by the binding of HH ligands to the transmembrane protein Patched 1 (PTCH1), which relieves its inhibition on the transmembrane protein Smoothened (SMO). As a consequence, active SMO triggers an intracellular signaling cascade leading to the formation of activating forms of the GLI zinc-finger transcription factors GLI2 and GLI3, which directly induce GLI1 transcription. GLI1 is a strong transcriptional activator, whereas GLI2 and GLI3 act as activators in their full-length forms or as repressors in their C-terminal cleaved forms [1–3]. All three GLIs recognize the consensus sequence 5′-GACCACCCA-3′ in the promoter of target genes [4].

Activation of the HH pathway in cancer cells can occur through the loss of the inhibitory function of PTCH1 [5] or by activating mutations in SMO [6]. Recent reports suggest that in addition to upstream HH signaling, GLI proteins can be modulated by other signaling pathways or by proliferative and oncogenic inputs [7]. Therefore, to understand the status of HH signaling in a cell or its modulation by a given factor or a drug, it is very informative to

Natalia A. Riobo (ed.), *Hedgehog Signaling Protocols*, Methods in Molecular Biology, vol. 1322, DOI 10.1007/978-1-4939-2772-2_7, © Springer Science+Business Media New York 2015

measure the transcriptional activity of HH pathway. Here, we describe luciferase reporter assays to study HH signaling transcriptional activity in normal and cancer cells. Briefly, cells are transfected with a reporter vector in which the expression of the luciferase gene is driven by a DNA sequence specifically recognized by the transcription factor (TF) of interest. Activation of the TF results in the binding to its response element, thus leading to the induction of the reporter. *Firefly* luciferase is a 61 kDa monomeric protein derived from *Photinus pyralis* that does not require posttranslational modifications to be functionally active [8, 9]. In presence of its substrate (beetle luciferin, ATP, magnesium, and molecular oxygen), it generates luminescence that can be measured by a luminometer in relative light units, before undergoing spontaneous inactivation. The luminescence generated is proportional to the amount of *Firefly* luciferase protein; therefore, changes in luminescence reflect changes in its transcription.

In a luciferase assay experiment, changes in the reporter response may be due to a specific effect on the reporter or to nonspecific effects, such as different transfection efficiency or differences in cell number. To distinguish between specific and nonspecific effects, the signal from the experimental reporter (*Firefly* luciferase) must be normalized to that of a control reporter (*Renilla* luciferase). The *Renilla* luciferase is driven by a constitutive promoter, often of viral origin, e.g., thymidine kinase (TK), cytomegalovirus (CMV), or Simian virus 40 (SV40), whose activity should not be affected by different experimental conditions. *Renilla* luciferase is a 36 kDa monomeric protein derived from *Renilla reniformis*, which is fully functional after translation and that becomes inactive after emitting luminescence [10]. As substrate, *Renilla* requires a coelenterate luciferin (coelenterazine) and molecular oxygen.

To measure the ability of GLI transcription factors to modulate gene expression, and thus to monitor the activity of HH pathway, a specific reporter (8×3′Gli-BS) has been developed by Sasaki and colleagues [11]. In this reporter vector, the transcription of the *Firefly* luciferase gene is driven by 8 tandem repeats of the Gli binding site (Gli-BS) followed by the δ-cristallin basal promoter. Activation of the 8×3′Gli-BS reporter can be triggered by canonical activation of the endogenous HH pathway (e.g., by Shh or SAG stimulation or by PTCH1 silencing) or by overexpression of the GLI transcription factors, thus providing the rationale to monitor the transcriptional activity of HH pathway at multiple levels. The luciferase assay with the 8×3′Gli-BS reporter, in fact, can be applied to (1) measure the activity of the endogenous HH pathway in different experimental conditions, (2) evaluate the effects of drugs or genes (by overexpression/knockdown) on endogenous HH pathway, (3) measure the activity of exogenous (overexpressed) GLI transcription factors or their variants (e.g., mutants, domain-deleted forms, etc.), and (4) evaluate the effects of drugs or genes (by overexpression/knockdown) on exogenous GLI transcription factors (Fig. 1a–d).

A luciferase reporter assay includes the following steps:

1. Seeding of the cells

2. Transfection of reporters and other vectors

3. Treatment with drugs (optional)

4. Reading of the luminescence

5. Data analysis

In the following sections, we will describe in detail all the steps, addressing the possible applications of the luciferase assay in the study of the transcriptional activity of the HH pathway in normal and cancer cells.

2 Materials

Described materials are for triplicate experiments in 12-well plates, but they could be scaled down to 24-, 48-, or 96-well plates.

2.1 Cell Culture

– Twelve-well plates for cell culture.

– Culture medium specific for the cell type used, with low percentage of fetal bovine serum (FBS) (≤ 1 %), if required.

2.2 Transfection

– Opti-MEM® (Life Technologies) or other serum-free medium suitable for transfection, depending on the method of transfection used.

– Transfection reagents (e.g., cationic lipid-based transfection reagents, calcium phosphate).

– Transfection-grade purified *Firefly* and *Renilla* reporters and any other vector encoding cDNA or short hairpin RNA (*shRNA*) to overexpress or silence, respectively, a factor of interest.

– Sterile 1.5 ml tubes to prepare the transfection mix.

– Culture medium specific for the type of cell used, with low percentage of fetal bovine serum (FBS) (≤ 1 %), if required.

2.3 Luminescence Reading

– Phosphate-buffered saline (PBS): 137 mM NaCl, 2.7 mM KCl, 10 mM Na_2HPO_4, 1.8 mM KH_2PO_4

– Dual-Luciferase® Reporter Assay System (Promega)

– Luminometer (e.g., GloMax® 20/20)

– 1.5 ml tubes

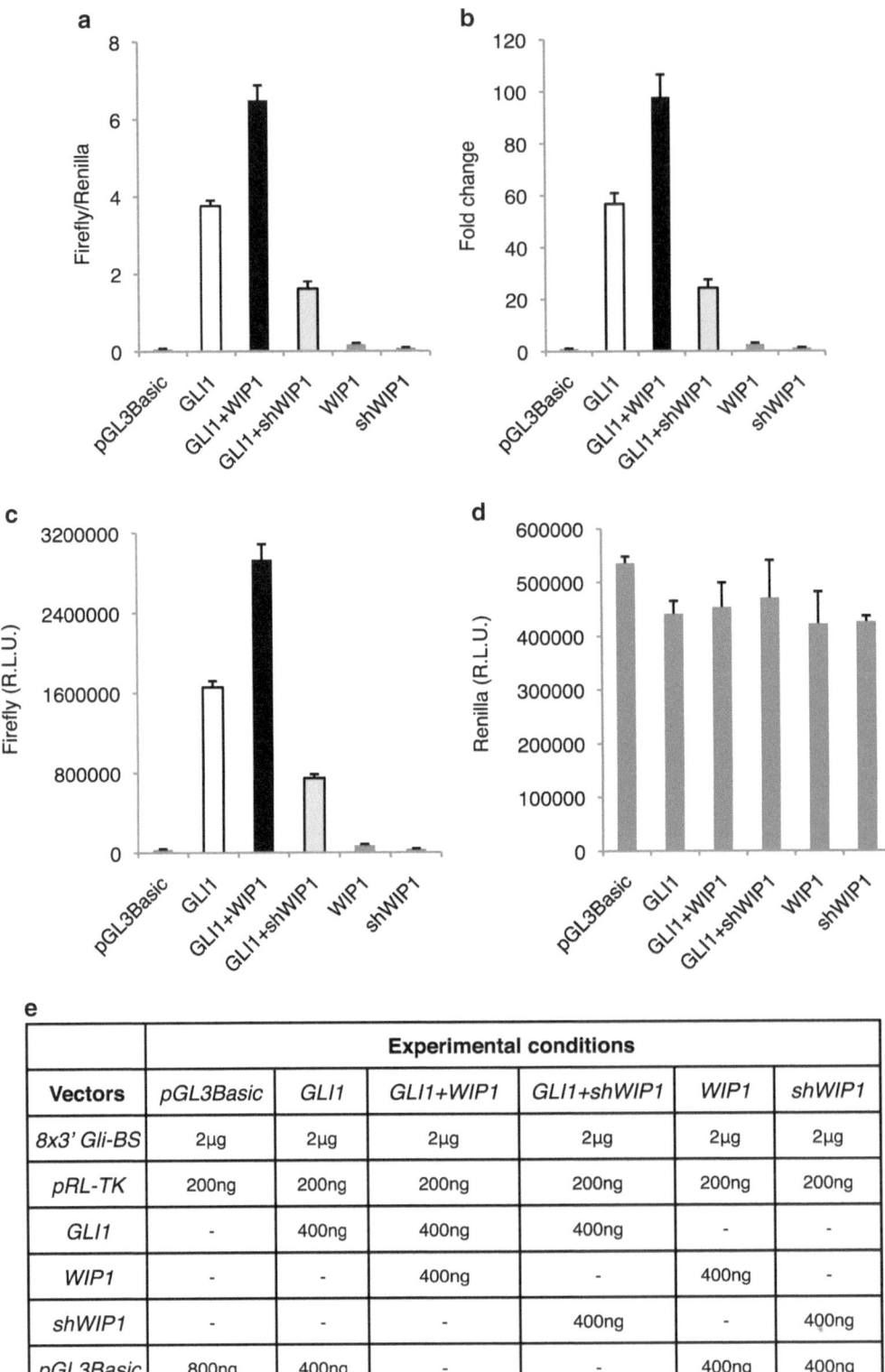

Fig. 1 Results of a luciferase assay in the HCT-116 colon cancer cell line. (**a**) Graph showing the Firefly/Renilla ratios in different experimental conditions. The graph shows that GLI1 transcriptional activity is increased by

3 Methods

3.1 Seeding of the Cells

The type and the number of cells depend on the experiment. To study the transcriptional activity of the endogenous HH pathway, cells responsive to HH stimulation or with an activated HH pathway should be used, such as:

1. NIH3T3 (ATCC CRL-1658) cells: mouse embryonic fibroblast cell line extensively used to study HH signaling at multiple levels. NIH3T3 cells readily activate HH signaling upon treatment with the ligand sonic hedgehog (SHH) or with the Smoothened agonist SAG [12].

2. Shh-LIGHT cells: a NIH3T3-derived cell line stably transfected with the $8 \times 3'$Gli-BS *Firefly* and pRL-TK *Renilla* luciferase reporters [13], ready to be used in luciferase assay.

3. SmoA1-LIGHT cells: a clonal subline of Shh-LIGHT cells expressing oncogenic Smo [13] and thus characterized by activated HH signaling.

4. Any type of normal or cancer cells that can be transfected with the reporters. However, not all cells are responsive to SHH or SAG treatment.

Cells must be seeded: (a) in their optimal growth medium, but with a reduced percentage of FBS (not higher than 1 %); (b) at a density that allows to get a confluency of about 50 % at the moment of transfection; and (c) in triplicate in 12-well plates (or scaled down to 24-, 48-, or 96-well plates).

3.2 Transfection of Reporters and Other Vectors

The total amount of DNA to be transfected in every single 12-well plate well is 1 μg, but because every condition must be assayed in triplicate, in the preparation of the mixture of transfection, the replicas should be taken into account. We found that the best reproducibility is obtained by preparing a transfection mixture sufficient for 3 wells (3 μg total).

Depending on the type of experiment, the following vectors should be used:

1. *Firefly* reporter: $8 \times 3'$ Gli-BS (see above).

2. *Renilla* reporter: pRL-TK that encodes the *Renilla* luciferase driven by the thymidine kinase promoter.

Fig. 1 (continued) overexpression of the oncogenic phosphatase WIP1 (GLI1 + WIP1), whereas it is decreased upon WIP1 silencing (GLI1 + shWIP1) [14]. (b) Graph showing fold change of Firefly/Renilla ratios, with the pGL3Basic equated to 1. (c, d) Graphs showing Firefly (c) and Renilla (d) luciferase values in different experimental conditions. Note that Renilla values do not change among different conditions. R.L.U., relative light units. Represented are mean values ± standard deviation. (e) Table describing the experimental conditions and the amount of transfected vectors, indicated as amount required for three wells of a 12-well plate

3. Vectors encoding GLI1, GLI2, or GLI3 and other factors of interest (pOVER) or specific shRNAs (pshRNA).

4. Empty vectors, such as pGL3Basic or similar, which contain the cDNA of the *Firefly* luciferase without a promoter to drive its expression. It is used as "filler" vector to equal the total amount of DNA among all conditions.

The combination of vectors to be transfected is critical. We optimized the assay by using the following equations to determine the relative amounts of each vector to be used in a single transfection mixture sufficient for a triplicate (3 wells of a 12-well plate, *see* Fig. 1e):

– Measurement of the transcriptional activity of endogenous HH pathway:

$$1x\,(\text{pRL-TK}) + 10x\,(8 \times 3'\text{Gli-BS}) = 3\ \mu g$$

– Measurement of the ability of a factor to modulate endogenous HH pathway:

$$1x\,(\text{pRL-TK}) + 10x\,(8 \times 3'\text{Gli-BS}) + 2x\,(\text{pOVER/} \\ \text{pshRNA}) = 3\ \mu g$$

– Measurement of the transcriptional activity of the GLI TFs:

$$1x\,(\text{pRL-TK}) + 10x\,(8 \times 3'\text{Gli-BS}) + 2x\,(\text{GLI1/2/3}) = 3\ \mu g$$

– Measurement of the transcriptional activity of the GLI TFs and their modulation by other genes:

$$1x\,(\text{pRL-TK}) + 10x\,(8 \times 3'\text{Gli-BS}) + 2x\,(\text{GLI1/2/3}) \\ + 2x\,(\text{pOVER/pshRNA}) = 3\ \mu g$$

– where x is the amount of DNA in micrograms

Example:

$$1x\,(\text{pRL-TK}) + 10x\,(8 \times 3'\text{Gli-BS}) + 2x\,(\text{GLI1}) \\ + 2x\,(\text{pshRNA}) = 3\ \mu g$$

$$1x + 10x + 2x + 2x = 15x$$

$$x = 3\ \mu g / 15 = 200\ \text{ng}$$

$$200\ \text{ng}\,(\text{pRL-TK}) + 2\ \mu g\,(8 \times 3'\text{Gli-BS}) + 400\ \text{ng}\,(\text{GLI1}) \\ + 400\ \text{ng}\,(\text{pshRNA})$$

The transfection method depends on the transfection efficiency of the cells and on their ability to survive transfection, as some cells may significantly suffer. We perform transfection by using cationic lipids. A general protocol is the following:

1. Prepare the transfection mixture of vectors to be transfected for each condition, sufficient for 3 wells.

2. Incubate the DNA with the transfection reagent in 300 µl of medium suitable for transfection (e.g., Opti-MEM®) for about 20 min.

3. Add 100 µl of the transfection mixture to the cells plus 400 µl of Opti-MEM, to have a final volume of 500 µl in each well.

4. After 6–8 h replace the transfection mix with 1 ml of the appropriate culture medium (≤1 % FBS).

3.3 Treatment with Drugs

3.3.1 Modulation of endogenous HH pathway

Activation of endogenous HH pathway in NIH3T3 and NIH3T3-derived cells can be easily achieved by treatment with the HH ligands, such as Shh (4 nM) [13], or with SMO agonists, such as SAG (100 nM) [*N*-methyl-*N'*-(3-pyridinylbenzyl)-*N'*-(3-chlorobenzo[b]thiophene-2-carbonyl)-1,4-diaminocyclohexane] [14]. The treatment should last 48 h before reading the luminescence; therefore, the ligand/agonist should be added to the cells the same day of the transfection when replacing the medium.

To investigate whether a drug affects the transcriptional activity of activated endogenous HH signaling, the drug should be added after the Shh ligand or SAG, so that the pathway can be activated before the drug is added. The duration of the treatment depends on the drug used and on its mechanism of action. If a drug is supposed to act at PTCH or SMO levels, the order of addition of the drug and of HH agonists can lead to different results that should be carefully interpreted.

3.3.2 Modulation of GLI Transcriptional Activity

If a drug is expected to affect at posttranscriptional level the ability of the GLI TFs to transactivate the reporter, it should be added after the transfection, so that the overexpressed proteins are produced.

3.4 Reading of the Luminescence

Approximately 48 h after transfection, the luminescence can be read. A widely used kit is the Dual-Luciferase® Reporter Assay System (Promega). The following is a typical protocol that uses GloMax® 20/20 luminometer (Promega) without injectors:

1. Remove and discard the culture medium. If cells grow in suspension, or if there is a significant amount of cells floating, collect them by centrifugation in 1.5 ml tubes and proceed to **step 3**.

2. Wash once with PBS.

3. Add to each well (or to each 1.5 ml tube, if cells have been collected by centrifugation) 250 µl of Passive Lysis Buffer 1X (*see* **Note 1**) and incubate at room temperature for 20 min with vigorous shaking (*see* **Note 2**).

4. Collect lysate and transfer it in a 1.5 ml tube. Keep the samples at room temperature.

5. For each sample, prepare 20 µl of Luciferase Assay Reagent II (LAR II) (*see* **Note 3**) in a clean tube.

6. Measure the *Firefly* luminescence by adding 2 µl of cell lysate to the tubes containing LAR II. Mix by pipetting (do not vortex) and read the luminescence. The luminometer should be

programmed to perform a 2-s premeasurement delay, followed by a 10-s measurement period for each reporter assay point.

7. After reading the *Firefly* luminescence, add to the same tube 20 μl of Stop & Glo® Reagent (*see* **Note 4**) and read the *Renilla* luciferase signal. Stop & Glo® Reagent quenches *Firefly* luciferase luminescence and concomitantly provides the substrate for the *Renilla* luciferase.

8. Record the values of *Firefly* and *Renilla* luciferases and their ratio.

3.5 Data Analysis

The data analysis of a luciferase assay performed with three technical replicas includes the following: (1) calculation of the average value and standard deviation of the *Firefly* luciferase signal for each triplicate (Firefly), (2) calculation of the average value and standard deviation of the *Renilla* luciferase signal for each triplicate (Renilla), (3) calculation of the average Firefly/Renilla ratio and of its standard deviation, and (4) calculation of the fold change of the Firefly/Renilla ratios of each sample relative to the control and of its standard deviation.

Data can be shown as histograms representing the Firefly/Renilla ratios or their fold change relative to the control (Fig. 1a, b). The calculation of the fold change of the Firefly/Renilla ratios allows an easy comparison among experiments or among biological replicas of the same experiments. It is extremely important to check the pattern of the average values of the *Firefly* and *Renilla* luciferase among the samples. In fact, some treatments or modulation of some genes can systematically affect the *Renilla* signal, which should be constant (Fig. 1d). As a consequence, a change in the Firefly/Renilla ratio, with a marked change in the *Renilla* signal and a weak/no change in the *Firefly* signal, may not be due to a specific effect on the Gli-dependent reporter, but to a side effect of a treatment or gene modulation on the *Renilla* signal. Of course, differences in the *Renilla* signal can also be due to differences in cell number at the moment of reading, and thus it is important to check whether evident differences in the number of cells are observed before proceeding with cell lysis.

4   Notes

1. Passive Lysis Buffer (PLB) 1X is obtained by adding 1 volume of PLB 5X provided by the Dual-Luciferase® Reporter Assay System to 4 volumes of distilled water.

2. Some cells such as NIH3T3 and NIH3T3-derived cells require a longer lysis in PLB 1X and 2–3 cycles of freeze and thaw before reading the luminescence. Scraping may also be required.

3. Luciferase Assay Reagent II (LAR II) is prepared by resuspending the Luciferase Assay Substrate in 10 ml of Luciferase Assay Buffer II provided by the Dual-Luciferase® Reporter Assay System. Once prepared, LAR II should be aliquoted and stored at −70 °C until use.

4. Stop & Glo® Reagent is prepared by adding 1 volume of Stop & Glo® Substrate 50X to 50 volumes of Stop & Glo® Buffer. It is best to prepare Stop & Glo® Reagent just before use.

Acknowledgments

This work was supported by grants from AIRC (Associazione Italiana per la Ricerca sul Cancro, projects IG-9566 and IG-14184) and Regional Health Research Program 2009. S.P. was supported by an AIRC fellowship.

References

1. Ingham PW, McMahon AP (2001) Hedgehog signaling in animal development: paradigms and principles. Genes Dev 15:3059–3087

2. Rohatgi R, Scott MP (2007) Patching the gaps in Hedgehog signalling. Nat Cell Biol 9: 1005–1009

3. Jiang J, Hui CC (2008) Hedgehog signaling in development and cancer. Dev Cell 15:801–812

4. Kinzler KW, Vogelstein B (1990) The GLI gene encodes a nuclear protein which binds specific sequences in the human genome. Mol Cell Biol 10:634–642

5. Hahn H, Wicking C, Zaphiropoulous PG, Gailani MR, Shanley S, Chidambaram A et al (1996) Mutations of the human homolog of Drosophila patched in the nevoid basal cell carcinoma syndrome. Cell 85:841–851

6. Xie J, Murone M, Luoh SM, Ryan A, Gu Q, Zhang C et al (1998) Activating Smoothened mutations in sporadic basal-cell carcinoma. Nature 391:90–92

7. Stecca B, Ruiz I, Altaba A (2010) Context-dependent regulation of the GLI code in cancer by HEDGEHOG and non-HEDGEHOG signals. J Mol Cell Biol 2:84–95

8. Wood KV, de Wet JR, Dewji N, DeLuca M (1984) Synthesis of active firefly luciferase by in vitro translation of RNA obtained from adult lanterns. Biochem Biophys Res Comm 124: 592–596

9. de Wet JR, Wood KV, Helinski DR, DeLuca M (1985) Cloning of firefly luciferase cDNA and the expression of active luciferase in Escherichia coli. Proc Natl Acad Sci U S A 82:7870–7873

10. Matthews JC, Hori K, Cormier MJ (1977) Purification and properties of Renilla reniformis luciferase. Biochemistry 16:85–91

11. Sasaki H, Hui C, Nakafuku M, Kondoh H (1997) A binding site for Gli proteins is essential for HNF-3beta floor plate enhancer activity in transgenics and can respond to Shh in vitro. Development 124:1313–1322

12. Chen JK, Taipale J, Young KE, Maiti T, Beachy PA (2002) Small molecule modulation of Smoothened activity. Proc Natl Acad Sci U S A 99:14071–14076

13. Taipale J, Chen JK, Cooper MK, Wang B, Mann RK, Milenkovic L et al (2000) Effects of oncogenic mutations in Smoothened and Patched can be reversed by cyclopamine. Nature 406:1005–1009

14. Pandolfi S, Montagnani V, Penachioni JY, Vinci MC, Olivito B, Borgognoni L, Stecca B (2013) WIP1 phosphatase modulates the Hedgehog signaling by enhancing GLI1 function. Oncogene 32:4737–4747

Chapter 8

Measuring Expression Levels of Endogenous *Gli* Genes by Immunoblotting and Real-Time PCR

Pawel Niewiadomski and Rajat Rohatgi

Abstract

Gli proteins are transcription factors that mediate the transcriptional effects of Hedgehog signaling in vertebrates. The activities of Gli2 and Gli3 are regulated primarily by posttranslational modifications, while Gli1 is mostly regulated at the transcriptional level. Detection of endogenous Gli proteins had been hampered by lack of good antibodies, but this problem has been mostly resolved in recent years. In this chapter we describe methods of detecting expression of endogenous *Gli* genes in whole-cell lysates and in subcellular fractions and also provide protocols for the measurement of Gli mRNA levels by quantitative real-time reverse transcriptase PCR (qPCR).

Key words Gli proteins, Immunoblotting, Real-time RT-PCR, Nuclear localization, Cell fractionation

1 Introduction

Gli proteins are the final effectors of the Hedgehog pathway in vertebrates. They are the homologues of the *Drosophila* protein cubitus interruptus [1]. Of the three mammalian Gli proteins, Gli1 is present at low levels in the absence of signal and its expression at the mRNA and protein level dramatically increases in cells exposed to a Hedgehog agonist. Instead of being regulated at the transcriptional level, Gli2 and Gli3 proteins are regulated by posttranslational modifications and subcellular trafficking [2–6]. Specifically, in the basal state, Gli2 and Gli3 are phosphorylated by protein kinase A at six conserved serine residues P1–P6 and subsequently by glycogen synthase kinase 3β and casein kinase 1 [7, 8]. These phosphorylation events lead to the inhibition of the transcriptional activator function of Gli2/3 [6] and to their processing by the proteasome into truncated repressor forms (Gli2/3R) or their complete proteasomal degradation [7, 9, 10]. When the Hedgehog pathway becomes activated, the phosphorylation of P1–P6 and proteasomal processing of Gli2/3 is abrogated, and instead Gli2/3 become phosphorylated at a distinct set of sites by an unknown kinase,

Natalia A. Riobo (ed.), *Hedgehog Signaling Protocols*, Methods in Molecular Biology, vol. 1322,
DOI 10.1007/978-1-4939-2772-2_8, © Springer Science+Business Media New York 2015

an event known as hyperphosphorylation [2–6]. Subsequently, they translocate to the tip of the primary cilium and to the nucleus, where they activate transcription [2–6, 11]. Consequently, while Gli1 expression can serve as a late, transcriptional readout of Hh signaling, Gli2 and Gli3 posttranslational modifications and subcellular localization changes can be detected <1 h after the induction of signaling [2, 3, 6]. In this protocol, we present methods to quantify Gli1 expression and assess the activation state of Gli2/3 using Western blotting.

2 Materials

2.1 Cell Culture Materials

1. NIH/3T3 cells (ATCC).

2. High-serum culture media: DMEM containing 10 % fetal bovine serum (FBS), 1× GlutaMAX, 1× nonessential amino acids, 1× sodium pyruvate, and 1× penicillin/streptomycin solution (all from Life Technologies).

3. Sterile phosphate-buffered saline (PBS), calcium- and magnesium-free.

4. 0.05 % trypsin-EDTA solution (Life Technologies or equivalent).

5. Low-serum culture media: same as high-serum media, except made with 0.5 % FBS instead of 10 %.

6. Cell culture dishes: Nunc Nunclon or equivalent tissue culture-treated 10-cm-diameter round dishes and 24-well plates or equivalent.

7. Bortezomib stock solution: 10-mM bortezomib (LC Laboratories) in DMSO, store at –20 °C.

8. Smoothened agonist (SAG) stock solution: 100-μM SAG (Axxora) in DMSO, store at –20 °C (see **Note 1**).

2.2 Materials for RNA Isolation and qPCR

1. RNAse removal solution (RNAseZap from Life Technologies or equivalent).

2. TRIzol (Life Technologies).

3. Chloroform.

4. RNAse-free water.

5. RNAse-free glycogen (UltraPure Glycogen from Life Technologies or equivalent).

6. 70 % ethanol solution in RNAse-free water.

7. 100 % isopropanol.

8. cDNA synthesis kit: iScript Supermix (Bio-Rad) or equivalent.

9. SYBR green qPCR mix: iTaq SYBR green Supermix (Bio-Rad) or equivalent (see **Note 2**).

10. qPCR plates and adhesive film (*see* **Note 3**).

11. Refrigerated tabletop microcentrifuge.

12. UV/vis spectrophotometer.

13. Thermocycler.

14. Real-time PCR instrument: Applied Biosystems 7900HT Fast or equivalent.

15. Primer stock solutions: 20-µM solutions of primers in nuclease-free water.

2.3 Materials for Cell Lysis and Protein Isolation

1. 1× PBS, calcium- and magnesium-free—precooled to 4 °C.

2. 10× RIPA salts: 500-mM Tris pH 7.4, 1.5-M NaCl.

3. RIPA buffer: 1× RIPA salts (50-mM Tris pH 7.4, 150-mM NaCl), 2 % Nonidet P-40 (or IGEPAL CA-630), 0.25 % sodium deoxycholate, 1-mM DTT, protease inhibitors (SIGMAFAST no EDTA or equivalent), and phosphatase inhibitors (1-mM NaF, 1-mM activated Na_3VO_4—*see* **Note 4**), 1-µM bortezomib (optional). Make fresh before each experiment.

4. Refrigerated tabletop microcentrifuge.

5. Protein concentration assay kit: Pierce BCA protein assay kit or equivalent (*see* **Note 5**).

2.4 Materials for Electrophoresis and Western Blotting

1. 4× Laemmli sample buffer: ready-made solution (Bio-Rad) or equivalent.

2. 1-M DTT in water: aliquot and store frozen at –20 °C.

3. 8 % SDS-PAGE gels.

4. Protein electrophoresis equipment and buffers—electrophoresis chamber, transfer apparatus, power supply, gel running and transfer buffers, and nitrocellulose.

5. Primary antibodies:

   • Rabbit anti-Gli1 (Cell Signaling, cat. # 2534S—use at 1:500 dilution)

   • Goat anti-Gli3 (R&D Systems, cat. # AF3635—use at 1:200 dilution)

   • Guinea pig anti-Gli2, made in house (*see* **Note 6**—use at 1 µg/mL)

   • Mouse anti-α-tubulin (Sigma, cat. # T-5326—use at 1:10,000 dilution)

   • Rabbit anti-lamin A (Abcam, cat. # ab26300—use at 1:1,000 dilution)

6. Reagents and equipment for chemiluminescence or fluorescence (Li-Cor Odyssey) detection, including appropriate blocking buffer and secondary antibodies.

2.5 Materials for Nuclear/ Cytoplasmic Fractionation

1. 1× PBS, calcium- and magnesium-free—precooled to 4 °C.

2. 10× triethanolamine/acetic acid (TEA/AA): mix 1.99-mL TEA, 0.859-mL AA, and 150-mL H_2O, and adjust pH to 7.4 with TEA.

3. 1-M HEPES pH 7.4.

4. Prepare buffers fresh before each experiment:

   - 10-mM HEPES: dilute 1-M HEPES with ice-cold water.

   - SEAT buffer: 1× TEA/AA, 250-mM sucrose, 1× SIGMAFAST protease inhibitors (no EDTA), 1-μM bortezomib, 1-mM NaF, and 1-mM activated Na_3VO_4.

   - 5× lysis buffer: 250-mM Tris, pH 7.4, 1.35-M NaCl, 5 % NP-40, 1-mM DTT, and 1-μM bortezomib.

   - Benzonase buffer: 1× TEA/AA, 250-mM sucrose, 1× SIGMAFAST protease inhibitors (no EDTA), 1-μM bortezomib, 1-mM NaF, 1-mM activated Na_3VO_4, 1-mM $MgCl_2$, and Benzonase (EMD Millipore) 20U/mL.

5. 1-mL syringes and 25-G needles.

6. Cell lifters (Sigma, cat. #CLS3008).

7. Cooled tabletop microcentrifuge.

8. Materials for SDS-PAGE as detailed in Subheading 2.3.

3 Methods

3.1 Cell Culture

1. Culture NIH/3T3 cells in high-serum culture media. Cells should be split before they reach 80 % confluency. Split every 2–3 days at a ratio of 1:5 to 1:8, as appropriate.

2. When the cells are ready to passage, rinse with sterile PBS and cover with 2 mL per 10-cm dish of pre-warmed trypsin-EDTA solution. Incubate at 37 °C for 3–5 min until the cells detach from the surface of the dish. Tap dish on the side to dislodge remaining cells and cover with 8 mL of high-serum media. Count cells or split at a specified ratio.

3. One to two days before the experiment, split cells into dishes or well plates:

   (a) Real-time·qPCR—one well from a 24-well culture plate per sample; a well from a 48- or 96-well plate also should suffice, but the quantity of RNA may be very low and RNA isolation may be problematic. Plate at 1.5×10^5 cells per well to obtain confluent cultures in 24–48 h.

   (b) Western blot from whole-cell lysates—one well from a 24-well culture plate. Plate at 1.5×10^5 cells per well to obtain confluent cultures in 24–48 h.

(c) Nuclear/cytoplasmic fractionation—10-cm dish is preferable. Plate at 4×10^6 cells per dish to obtain confluent cultures in 24–48 h.

4. After the cells have reached full confluence (look "overcrowded"), change the media to low-serum culture media for 24–40 h (*see* **Note 7**).

5. Add Hedgehog-modulating drugs as needed for 2–24 h before harvesting (*see* **Note 8**).

3.2 RNA Isolation and qPCR

RNAse contamination must be avoided at all steps. Use dedicated RNAse-free sterile microcentrifuge tubes and pipette tips and RNAse-free solutions. Wipe all work surfaces with RNAseZap or equivalent RNAse removal solution. Always wear fresh powder-free gloves—do not reuse. Avoid touching surfaces that may have had contact with the skin or hair, such as chairs, desks, door handles, etc., with your gloves. All procedures involving TRIzol and chloroform should be performed in a chemical safety cabinet. TRIzol and chloroform leftovers should be disposed as hazardous waste.

1. Rinse cells with ice-cold PBS. Remove PBS carefully.

2. Add 400-μL TRIzol per well in 24-well plates. Pipette up and down several times to thoroughly lyse cells, and leave for 5 min at room temperature.

3. Transfer the lysate to 1.5-mL microcentrifuge tubes. Add 80-μL chloroform.

4. Shake vigorously for 15 s and incubate for 2 min at room temperature.

5. Centrifuge at $12,000 \times g$ for 15 min at 4 °C.

6. Remove upper aqueous phase to a separate tube (*see* **Note 9**). Discard the red organic phase and the interface as hazardous waste.

7. Add 5 μg of RNAse-free glycogen to the aqueous layer.

8. Add 200-μL isopropanol and mix by vortexing.

9. Centrifuge at $12,000 \times g$ for 15 min at 4 °C.

10. Decant supernatant carefully but thoroughly to avoid dislodging the pellet. Add 1 mL of 70 % ethanol. Mix gently and centrifuge at $7,500 \times g$ for 10 min.

11. Decant supernatant and remove remaining droplets by pipetting or aspirating with a capillary tip. Take care to not touch the pellet.

12. Air-dry the pellet (*see* **Note 10**).

13. Dissolve the pellet in 20-μL RNAse-free water by pipetting up and down around 20 times (*see* **Note 11**).

Table 1
Primers for qPCR reactions

Primer	Sequence	Concentration (nM)
GAPDH forward	GGCCTTCCGTGTTCCTAC	100
GAPDH reverse	TGTCATCATACTTGGCAGGTT	100
Gli1 forward	CCAAGCCAACTTTATGTCAGGG	200
Gli1 reverse	AGCCCGCTTCTTTGTTAATTTGA	200
Gli2 forward	GTGCACAGCAGCCCCACACTCTC	50
Gli2 reverse	GGTAATAGTCTGAAGGG TTGGTGCCTGG	50

14. Measure RNA concentration of samples using a UV/vis spectrophotometer.

15. Make cDNA using iScript Supermix according to the manufacturer's instructions. Use 0.5-µg RNA per 20-µL reaction (*see* **Note 12**). Freeze the remaining RNA and store at −80 °C.

16. Perform qPCR according to the instructions of your qPCR instrument. Use the following parameters:

    (a) Primer concentration 50–200 µM (*see* Table 1)

    (b) Annealing temperature 55–60 °C

    (c) Annealing time 30–45 s

    (d) Extension time 10–15 s

    (e) 1 µL of the cDNA reaction/well

3.3 Detection of Endogenous Gli Proteins in Whole-Cell Lysates by Western Blot

1. Precool the centrifuge to 4 °C.

2. Remove media from cells and rinse briefly twice in ice-cold PBS.

3. Add 200-µL ice-cold RIPA buffer per well in a 24-well plate. Pipette up and down several times. Shake plate vigorously at 4 °C for 15–30 min to ensure complete lysis. If using larger culture surfaces, cells can be scraped off in lysis buffer and transferred to 1.5-mL microcentrifuge tubes prior to shaking.

4. Transfer lysate to a 1.5-mL microcentrifuge tube. Centrifuge at $21,000 \times g$ for 30 min at 4 °C.

5. Move supernatant to a fresh tube and keep on ice.

6. Measure protein concentration using the protein concentration assay kit.

7. Transfer equal quantities of protein (20–50 μg/sample) to fresh tubes and add 4× sample buffer and DTT to 50 mM.

8. Denature protein by heating at 65 °C for 15 min in a heat block.

9. Centrifuge at room temperature for 5 min at 21,000×*g*. Load samples on gel.

10. Run on 8 % SDS-PAGE gel at 120 V for approximately 2 h, until the dye front reaches the bottom of the gel. Transfer to nitrocellulose using wet transfer or Trans-Blot Turbo (Bio-Rad; *see* **Note 13**).

11. Detect by Western blot using chemiluminescence or fluorescence (Li-Cor Odyssey). Overnight incubation at 4 °C with primary antibodies is recommended, especially for anti-Gli3. *See* Fig. 1 for example blots (*see* **Note 14**).

3.4 Detection of Gli2/3 Nuclear Translocation and Hyperphosphorylation by Subcellular Fractionation

Perform all manipulations on ice or in the cold room at 4 °C.

1. Precool the centrifuge to 4 °C.

2. Remove media from a 10-cm dish of cells and rinse briefly twice in ice-cold PBS. Remove PBS carefully.

3. Rinse cells in ice-cold 10-mM HEPES pH 7.4. Remove the solution carefully.

4. Cover cells with ice-cold 10-mM HEPES pH 7.4 and leave on ice for 10 min.

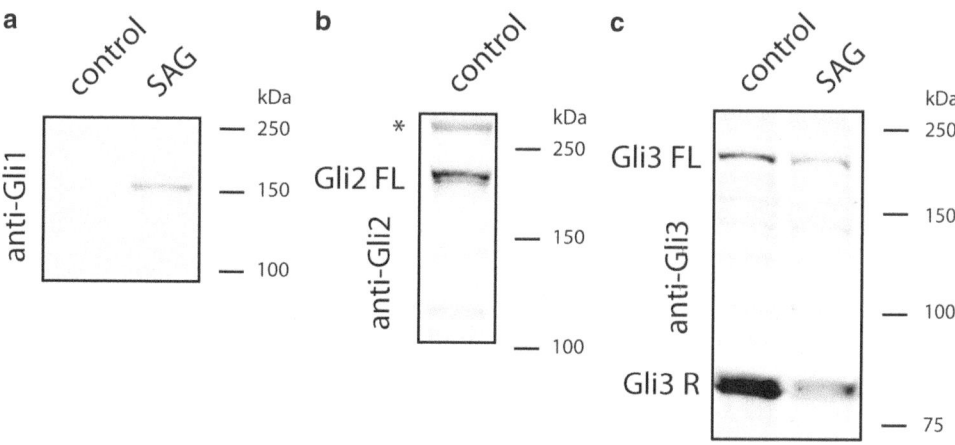

Fig. 1 Detection of endogenous Gli proteins in whole-cell lysates by immunoblotting. IrDye-coupled secondary antibodies were used, and the membranes were scanned using the Li-Cor Odyssey system. (**a**) NIH/3T3 cells were either left untreated or treated for 12 h with 100 nM SAG. Gli1 levels were measured by anti-Gli1 immunoblotting. (**b**) Anti-Gli2 immunoblot of NIH/3T3 lysate. *nonspecific band; Gli2 FL, full-length Gli2. Gli2 repressor is not detected by our antibody. (**c**) NIH/3T3 cells were treated as in **a**. Anti-Gli3 immunoblot of whole-cell lysate is shown. Gli3 FL, full-length Gli3; Gli3 R, truncated repressor form of Gli3. Please note reduced levels of Gli3 R in cells treated with SAG

5. Carefully remove all of the HEPES solution.

6. Add 500 µL of ice-cold SEAT buffer. Swirl the plate to distribute the SEAT solution over the cells.

7. Scrape cells using a cell lifter and transfer into 1.5-mL micro-centrifuge tube.

8. Bring the volume up to 750 µL with SEAT buffer.

9. Shear cells with a syringe attached to a 25-G needle. For each sample make 15 up-and-down strokes. Take care to not introduce bubbles.

10. Centrifuge at $900 \times g$ for 5 min. at 4 °C.

11. Transfer supernatant (600 µL) to a fresh tube and place on ice (cytoplasmic or "C" fraction). Resuspend pellet (nuclear or "N" fraction) in 1 mL of SEAT buffer by pipetting up and down with a 1-mL tip.

12. Centrifuge both C and N fractions at $900 \times g$ for 5 min. at 4 °C.

13. For the N fraction:

    (a) Remove supernatant carefully without disturbing the pellet. Resuspend the pellet in 550 µL of Benzonase buffer and incubate on ice for 10 min.

    (b) Add 137.5 µL of 5× lysis buffer and pipette up and down to mix.

14. For the C fraction.

    (a) Transfer supernatant (550 µL) to a new tube. Discard the pellet.

    (b) Add 137.5 µL of 5× lysis buffer and pipette up and down to mix.

15. Lyse both fractions with shaking or rotation at 4 °C for 45 min.

16. Centrifuge at $21,000 \times g$ for 30 min at 4 °C. Transfer supernatants to new tubes.

17. Optional—to increase sample concentration precipitate sample by the chloroform/methanol method or the trichloroacetic acid method. Resuspend the pellet in 1× SDS sample buffer + 50-mM DTT for 30 min. at 37 °C with vigorous mixing.

18. Transfer 70 µL of each fraction into a separate tube. Add 25 µL of 4× Laemmli sample buffer and 5 µL of 1-M DTT. Denature protein at 65 °C for 15 min in a heat block (*see* **Note 15**).

19. Detect Gli proteins by electrophoresis/immunoblotting as in Subheading 3.3 (*see* **Note 16**). To control for efficiency of separation between fractions, use anti-α-tubulin (cytoplasmic) and anti-lamin A (nuclear) antibodies as loading controls. An example immunoblot is shown in Fig. 2.

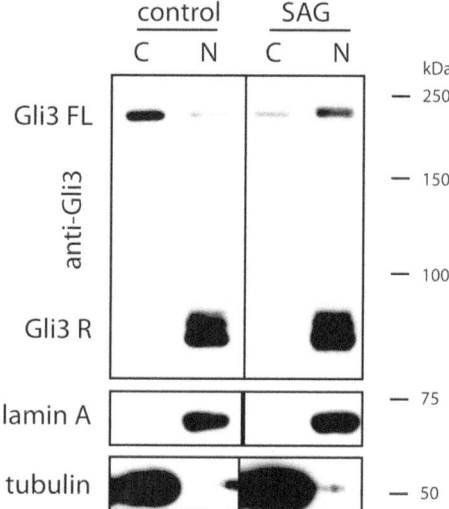

Fig. 2 Nuclear translocation of Gli3 upon Hh pathway activation. NIH/3T3 cells were treated for 2 h with 100-nM SAG and subjected to subcellular fractionation. Cytoplasmic and nuclear fractions were resolved by electrophoresis and immunoblotted with anti-Gli3, anti-lamin A, and anti-tubulin. HRP-coupled secondary antibodies and chemiluminescence-based detection were used. Please note the increased amount of full-length Gli3 (Gli3 FL) in the nuclear fraction of treated cells. Gli3 in the nuclear fraction of SAG-treated cells runs slightly higher on gel than in untreated cells, which reflects hyperphosphorylation of activated Gli3. Gli3 repressor (Gli3 R) is constitutively located at the nucleus. Lamin A was used as a nuclear marker and tubulin as cytoplasmic marker

4   Notes

1. SAG has a tendency to precipitate in aqueous solutions. When adding it to the cell culture, first dilute the stock solution in 1–2 mL of media and immediately vortex for 5–10 s, and then add the diluted solution to a dish.

2. Real-time qPCR master mix must be compatible with the available qPCR instrument. Check compatibility with your instrument on the manufacturer's website. We used iTaq SYBR Green Supermix with ROX (Bio-Rad, cat. #172-5851) with the Applied Biosystems 7900HT Fast system.

3. Plates and adhesive film must be tested for compatibility with your qPCR instrument. Contact the manufacturer for a list of compatible consumables. We used 384-well clear optical reaction plates (Applied Biosystems, cat. #4309849) and optical adhesive film (Applied Biosystems, cat. #4311971) with the Applied Biosystems 7900HT Fast equipped with a 384-well plate adapter. However, many systems are only compatible with 96-well plates.

4. Sodium orthovanadate must be activated prior to use. Prepare a 200-mM solution and adjust the pH to 10.0. Boil the solution until it loses its yellow color. Cool to room temperature. Readjust the pH to 10.0 and repeat until the solution remains colorless and the pH stabilizes at 10.0. Store aliquoted at –20 °C.

5. The BCA assay is sensitive to the presence of reducing agents such as DTT in the sample. The lysis buffer contains 1-mM DTT. To avoid errors in protein concentration measurement, add to all standard curve samples (including the negative control) a volume of lysis buffer equal to the volume of sample used for protein concentration measurement. For instance, if you add 2 μL of sample to each assay, you need to add 2 μL of lysis buffer to each of the standard curve assays and the negative control assay. If protein concentration cannot be measured immediately after lysis, the leftover lysis buffer should be stored with the samples and used in the protein concentration assay to account for DTT degradation over time.

6. The anti-Gli2 antibody we use for immunoblotting of Gli2 was custom made in guinea pigs based on Cho et al. [12]. The antigen was a his-tagged fragment containing amino acids 1053–1264 of mouse Gli2 (VQYIKAH… AKPSHLG), and the sera were affinity purified on amino-link beads coupled to the same antigen.

7. Longer serum starvation times accelerate the response and are therefore preferable in experiments where shorter drug treatments are applied (nuclear/cytoplasmic fractionation, qPCR). For 24-h treatments, serum starvation may be started at the same time as treatment.

8. Nuclear translocation and hyperphosphorylation of Gli2/3 becomes visible after 1–3 h of treatment with 100-nM SAG. Longer treatment times result in degradation of full-length Gli3 [2]. After 2–3 h of treatment, a clear increase in Gli1 mRNA becomes apparent in qPCR assays. Gli1 protein levels rise after 6–8 h of SAG treatment.

9. Take care not to touch the interface, which contains genomic DNA. Around 180 μL of aqueous phase can be withdrawn relatively safely without touching the interface.

10. The pellet should become transparent as it dries. Dissolve the pellet in water as soon as it becomes fully transparent. Do not allow the pellet to dry for too long, since it becomes more difficult to resuspend when completely dry.

11. Do not use DEPC-treated water for the resuspension of RNA pellet, since ethanol traces in the solution may interfere with subsequent reactions.

12. Using 1 μg of RNA per reaction, as recommended by the manufacturer, may result in nonlinear amplification results for the most highly expressed genes, such as GAPDH.

13. Standard-voltage semidry transfer results in poor transfer of high molecular weight proteins, such as Gli proteins. High-voltage semidry transfer (Trans-Blot Turbo) or wet transfer results in significantly better Gli protein signal on the membrane. Nitrocellulose and PVDF can both be used as membrane material, but PVDF gives high background when fluorescence scanner (Li-Cor Odyssey) is used for the detection.

14. Gli1 can be detected on the same membrane as Gli2 or Gli3 due to differences in the apparent molecular weight (Gli1 ~150 kDa, Gli2 and Gli3 full-length ~200 kDa, Gli3 repressor ~100 kDa). Cut the membrane into strips at 165 kDa and 120 kDa and blot the top and bottom strip with anti-Gli2 or anti-Gli3 and the middle strip with anti-Gli1. Anti-Gli2 was developed against the C-terminal part of the full-length protein and will not detect Gli2 repressor.

15. Do NOT attempt to normalize sample volumes by measuring protein concentrations. Cytoplasmic and nuclear fractions naturally have different concentrations of proteins. The method relies on careful manipulation to ensure that sample loss throughout the procedure is minimal and consistent between samples and fractions, so different samples can typically be directly compared to one another. Cytoplasmic and nuclear levels of protein can be normalized after immunoblotting to levels of loading control proteins: tubulin for cytoplasmic and lamin A for nuclear fraction.

16. After nuclear/cytoplasmic fractionation, the samples are relatively dilute. To enhance detection of low abundance proteins, such as Glis, it is recommended to concentrate the sample by precipitation. Precipitation has the added advantage of desalting the sample and making it run somewhat better on gel. High-sensitivity chemiluminescence reagents (SuperSignal West Femto—Pierce or VisiGlo Select—Amresco) may need to be used to detect Gli2/3 in dilute samples. Alternatively, fluorescence detection using the Li-Cor Odyssey system has proved in our hands to have sufficient sensitivity to detect Gli proteins, even in nuclear translocation assays.

Acknowledgments

This work was supported by the OPUS grant from the Polish National Science Centre (grant 2014/13/B/NZ3/00909) to P.N., a grant from the US National Institutes of Health (R21 NS074091) to R.R. , a Distinguished Scientist Award from the Sontag Foundation to R.R., a Scholar award from the Pew Foundation to R.R., and a Scholar award from the V Foundation for Cancer Research to R. R..

References

1. Hui C-C, Angers S (2011) Gli proteins in development and disease. Annu Rev Cell Dev Biol 27:513–537

2. Humke EW, Dorn KV, Milenkovic L, Scott MP, Rohatgi R (2010) The output of Hedgehog signaling is controlled by the dynamic association between Suppressor of Fused and the Gli proteins. Genes Dev 24:670–682

3. Tukachinsky H, Lopez LV, Salic A (2010) A mechanism for vertebrate Hedgehog signaling: recruitment to cilia and dissociation of SuFu-Gli protein complexes. J Cell Biol 191: 415–428

4. Haycraft CJ, Banizs B, Aydin-Son Y, Zhang Q, Michaud EJ, Yoder BK (2005) Gli2 and Gli3 Localize to Cilia and Require the Intraflagellar Transport Protein Polaris for Processing and Function. PLoS Genet 1:e53

5. Niewiadomski P, Zhujiang A, Youssef M, Waschek JA (2013) Interaction of PACAP with Sonic hedgehog reveals complex regulation of the hedgehog pathway by PKA. Cell Signal 25: 2222–2230

6. Niewiadomski P, Kong JH, Ahrends R, Ma Y, Humke EW, Khan S et al (2014) Gli protein activity is controlled by multisite phosphorylation in vertebrate Hedgehog signaling. Cell Rep 6:168–181

7. Pan Y, Bai CB, Joyner AL, Wang B (2006) Sonic hedgehog signaling regulates Gli2 transcriptional activity by suppressing its processing and degradation. Mol Cell Biol 26:3365–3377

8. Price MA, Kalderon D (1999) Proteolysis of cubitus interruptus in Drosophila requires phosphorylation by protein kinase A. Dev Camb Engl 126:4331–4339

9. Tempe D, Casas M, Karaz S, Blanchet-Tournier MF, Concordet JP (2006) Multisite Protein Kinase A and Glycogen Synthase Kinase 3{beta} Phosphorylation Leads to Gli3 Ubiquitination by SCF{beta}TrCP. Mol Cell Biol 26: 4316–4326

10. Pan Y, Wang C, Wang B (2009) Phosphorylation of Gli2 by protein kinase A is required for Gli2 processing and degradation and the Sonic Hedgehog-regulated mouse development. Dev Biol 326:177–189

11. Kim J, Kato M, Beachy PA (2009) Gli2 trafficking links Hedgehog-dependent activation of Smoothened in the primary cilium to transcriptional activation in the nucleus. Proc Natl Acad Sci U S A 106:21666–21671

12. Cho A, Ko HW, Eggenschwiler JT (2008) FKBP8 cell-autonomously controls neural tube patterning through a Gli2- and Kif3a-dependent mechanism. Dev Biol 321:27–39

Chapter 9

Quantitative Immunoblotting of Endogenous Hedgehog Pathway Components

Shohreh F. Farzan and David J. Robbins

Abstract

Quantitative analysis and modeling of signaling pathway components can reveal important information about the dynamics of that system, including the relative stoichiometries and affinities between the individual signaling components, as well as rate-limiting steps in the signaling pathway. In this chapter, we present a method that we developed to quantify the steady-state ratio of core Hedgehog (Hh) signaling components in both cultured cells and the *Drosophila* embryo, a physiologically relevant tissue.

Key words Quantification, Quantitative immunoblotting, Hedgehog, Stoichiometry, Costal-2, Fused, Cubitus interruptus, Smoothened, Suppressor of Fused, *Drosophila*

1 Introduction

Consider a loaf of bread. When water, flour, yeast, and salt are combined in the correct proportions, a dietary staple emerges from the oven. However, alter the amount of any single ingredient, and one is often left with something inedible. Whether one is baking bread or constructing a model of Hedgehog (Hh) signaling, the importance of knowing the correct ratio of components is undeniable, if somewhat forgettable. Very different interpretations of the same set of data can emerge when one considers the stoichiometric ratios or proportions of the various signaling components (Fig. 1). As described in previous work [1], our group determined the endogenous, steady-state concentration of five core components: Smoothened (Smo), Cubitus interruptus (Ci), Costal-2 (Cos2), Fused (Fu), and Suppressor of Fused (Sufu), within a typical *Drosophila* Clone-8 (Cl-8) cell lysate (Fig. 1). In this cultured cell line and in validation experiments with *Drosophila* S2 cells and embryo lysates, we determined that Smo is present in limiting amounts in comparison to all other components and that Sufu is present in vast excess to these other signaling components [1].

Natalia A. Riobo (ed.), *Hedgehog Signaling Protocols*, Methods in Molecular Biology, vol. 1322,
DOI 10.1007/978-1-4939-2772-2_9, © Springer Science+Business Media New York 2015

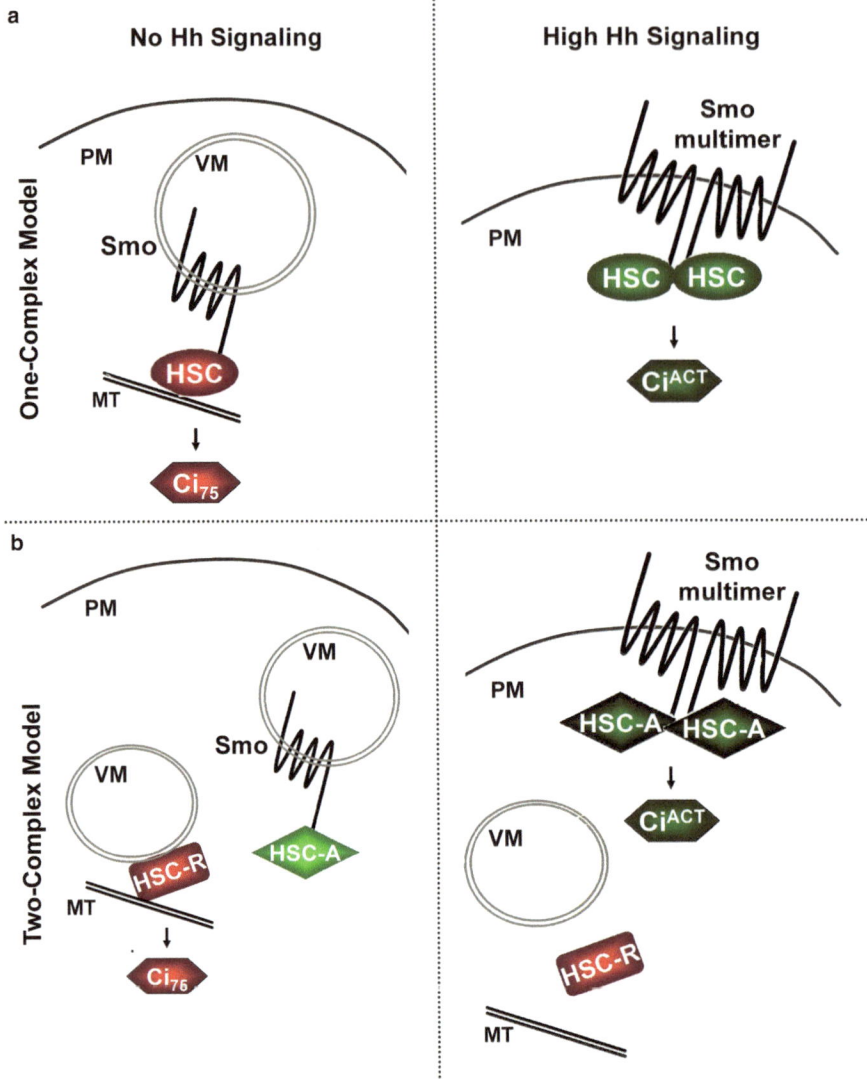

Fig. 1 Models of Hh signaling. A number of models have been proposed for how the Hh signaling components localize and interact to transduce the Hh signal. These various models can be simplified into two distinct signaling strategies, one in which all of the signaling components are bound directly to Smo in a membrane-associated complex that regulates all Ci activity (the one-complex model) and one in which two forms of the complex exist to differentially regulate the repressor and activator functions of Ci (the two-complex model). One major prediction of the two-complex model is that these various Hh signaling components will not exist in equimolar ratios within cells, with Smo existing in limiting amounts relative to Cos2. We used the method presented in this chapter to begin to test the predictions of these models by quantifying the molar ratio of the core downstream Hh signaling components. We showed that Smo is indeed a limiting component within cells, whose steady-state molar ratio is approximately one-tenth that of Cos2 and Fu. These models and the results of our quantification are explained in greater detail in Farzan et al. [1]. Abbreviations: plasma membrane (PM), vesicular membranes (VM), microtubules (MT), and Hedgehog signaling complex (HSC) consisting of Fu, Cos2, and Ci. The various activities of components are indicated by color (*red*, repressor; *green*, activator), and Ci^{ACT} refers to a highly activated state. Sufu (not depicted) is predicted to be cytosolic by these models. (Figure reproduced with author's permission from Farzan et al. Journal of Biological Chemistry. [1])

Similarly, quantitation of Wnt pathway members identified that low levels of Axin act in a rate-limiting manner and provided insight into fluctuations in the β-catenin response, further demonstrating the importance of this technique [2]. Together, these results demonstrate that quantitation of signaling components can provide the first step toward developing more dynamic models of component interactions and can expand the mechanistic insight into how signal transduction pathways operate within cells [1–3].

Here, we outline the quantitative immunoblotting technique that we developed as a general method that can be used to quantitate any component of interest. Our technique, which utilizes basic methods and standard equipment, has the potential to be adapted for a variety of purposes, from quantifying relative amounts of additional Hh signaling components, and measuring their concentrations in additional cell types and tissues of interest, to providing quantitative information about other signal transduction pathways. As the methods for generating purified proteins are beyond the scope of this chapter and protein purification techniques have been previously published, we refer the reader to other sources for further information on generating epitope-tagged purified Hh components. For more details on the purified proteins used in our method, please *see* **Note 1** and references [1, 4–7]. We will begin by assuming that the reader has generated a stock of a purified recombinant Hh component that is of unknown concentration (referred to as "Hh component X") and wishes to know the endogenous concentration of Hh component X in a cell lysate. As most readers will also be familiar with standard techniques for sodium dodecyl sulfate polyacrylamide gel electrophoresis (SDS-PAGE) and immunoblotting, we will focus on details of quantitation by immunoblotting.

2 Materials

2.1 Cell Culture Components

For our assays, we quantified the amount of Hh components in Cl-8 insect cells and *Drosophila* embryo lysates. Lysates of other tissues of interest can be substituted, depending upon the research question.

1. Cl-8 cell line stocks.

2. Cl-8 cell culture medium: Shields and Sang insect medium (Sigma), supplemented with 2 % FBS, 2.5 % fly extract, 0.0125 IU/ml insulin, 1 % penicillin-streptomycin.

3. *Drosophila* embryos: See Schafer et al. for details of large-scale embryo production and harvest [8].

2.2 Components for Making Cell Lysates and Subcellular Fractionation

For more details on subcellular fractionation, please refer to Stegman and Robbins [9].

1. Glass Dounce homogenizer.

2. Hypotonic lysis buffer (HLB): 50 mM β-glycerophosphate, 10 mM NaF, 1.5 mM EGTA, 1 mM DTT, pH 7.6.

3. Membrane-enriched pellet wash buffer: HLB supplemented with 150 mM NaCl.

4. Membrane-enriched pellet resuspension buffer: HLB supplemented with 1 % Nonidet P-40 (NP-40).

5. Embryo lysis buffer: 1 % NP-40 lysis buffer (1 % NP-40, 150 mM NaCl, 50 mM Tris, 50 mM NaF, pH 8.0).

2.3 Components for Staining and Quantification of Purified Protein Standards

For all SDS-PAGE, we used standard techniques and equipment to cast gels and separate proteins under denaturing conditions, with consideration for the molecular weights of the component protein of interest and the bovine serum albumin (BSA) standard protein when selecting the percentage of acrylamide to be used.

1. SDS stacking/resolving gel.

2. SDS-PAGE running buffer.

3. Purified bovine serum albumin (BSA) standard (Thermo Scientific).

4. Laemmli loading buffer [10].

5. Coomassie stain [10].

6. Silver stain [11].

7. SYPRO Ruby (Life Technologies) or other in-gel protein stains (optional).

2.4 Components for Quantification of Endogenous Hh Component X

1. SDS stacking/resolving gel ingredients.

2. SDS-PAGE running buffer.

3. Nitrocellulose or PVDF transfer membrane.

4. Transfer apparatus and blotting materials.

5. SDS-PAGE transfer buffer.

6. Membrane blocking and washing buffers.

7. Primary antibody against "Hh component X" and corresponding secondary antibody.

8. ECL Plus or other immunoblotting substrates that can be detected with fluorescent imaging equipment.

9. Fluorescent imaging equipment (e.g., STORM imaging system) and image analysis software (e.g., ImageQuant).

3 Methods

3.1 Quantify Amount of Recombinant Purified Hh Component X Against a BSA Standard Curve

1. Prepare two duplicate gels for SDS-PAGE. When pouring the stacking gels, select combs with ~15 wells to ensure there are enough wells to accommodate the molecular weight standard, the BSA standard curve, and two to three different volumes of the purified component X in duplicate.

2. As the gels are polymerizing, prepare the protein samples for gel electrophoresis. Prepare purified component X by adding Laemmli loading buffer (e.g., add 1 volume purified component X plus 1 volume 2× Laemmli loading buffer). Boil protein samples for 5 min. Be sure that you have prepared an adequate volume for two gels.

3. Prepare a range of dilutions of BSA, using the same buffer as that in which purified component X is in to dilute the BSA stock. Begin by diluting the BSA stock to make eight different BSA dilutions equal to 2 ng/μl, 4 ng/μl, 8 ng/μl, 12 ng/μl, 16 ng/μl, 20 ng/μl, 30 ng/μl, and 40 ng/μl. Dilute stock dilutions 1:1 with 2× Laemmli loading buffer and boil protein samples for 5 min. Final BSA dilution concentrations will be 1 ng/μl, 2 ng/μl, 4 ng/μl, 6 ng/μl, 8 ng/μl, 10 ng/μl, 15 ng/μl, and 20 ng/μl.

4. Transfer the polymerized gel to the electrophoresis apparatus, fill the chambers with running buffer, and load the protein samples onto the gel with the protein samples (Fig. 2a). In the first lane, load a molecular weight standard. In lanes 2 through 9, load 10 μl of each BSA dilution to generate the standard curve. When 10 μl of each dilution is loaded onto the gel, the final on-gel concentrations of the BSA standards will be 10 ng, 20 ng, 40 ng, 60 ng, 80 ng, 100 ng, 150 ng, and 200 ng.

5. In the remaining lanes, load a range of volumes of Hh component X sample (Fig. 2a). For example, load 5 μl in both lanes 10 and 11, 10 μl in both lanes 12 and 13, and 15 μl in lanes 14 and 15. Record the volumes of Hh component X loaded in each well. It may be necessary to adjust the amounts loaded onto the gel depending upon the estimated concentration of Hh component X sample to keep it in the range of the BSA standard curve (*see* **Note 2**). Repeat **steps 4** and **5** to load the second gel.

6. Run the gels until the molecular weight marker indicates that purified component X and BSA standards have been sufficiently resolved. Remove the gels from the electrophoresis chamber and transfer one gel to a clean, glass staining tray for silver staining and the other gel to a second, clean staining tray for Coomassie staining.

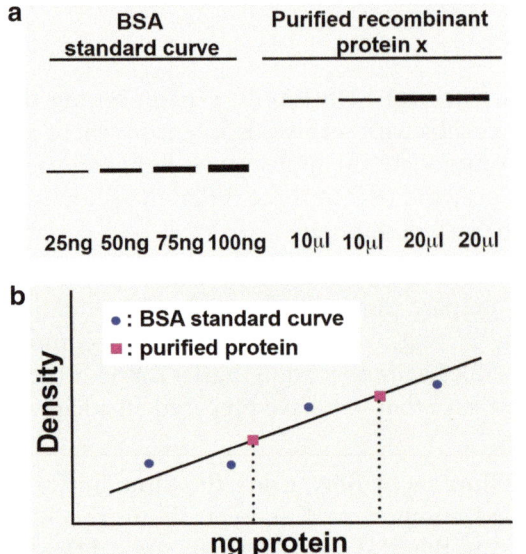

a

BSA
standard curve   Purified recombinant
                              protein x

25ng 50ng 75ng 100ng   10μl 10μl 20μl 20μl

b

• : BSA standard curve
■ : purified protein

Density

ng protein

1. **Run purified recombinant Hh component protein X against a BSA standard curve**

2. **Stain proteins to visualize in gel**

3. **Quantify density values with image analysis software**

4. **Quantify purified recombinant Hh component protein X against BSA standard curve values**

Fig. 2 Quantification of recombinant purified Hh component X against a BSA standard curve. (**a**) Different volumes of purified recombinant Hh component X and a standard curve of BSA are run on an SDS-PAGE gel and stained. (**b**) The densities of stained bands are quantified, and the concentration of purified Hh component X is determined using the BSA standard curve

7. Stain gel 1 using silver staining protocol described in reference [11] to reveal protein bands (Fig. 2a).

8. Stain gel 2 using Coomassie staining protocol described in reference [10] to reveal protein bands (Fig. 2a).

9. Scan stained gel 1 to generate a JPG or TIFF image file for each gel that can be quantified with image analysis software, such as ImageQuant (GE Healthcare). For most scanners, you may first insert the gel into a clear plastic sleeve or a piece of plastic film—with all bubbles and wrinkles removed prior to scanning—to protect the scanner.

10. Use ImageQuant (or similar software) to quantify the density of each protein band individually. Density values are usually generated by drawing boxes of equal size around each band to be quantified. The software then analyzes the density of the image within each box, and the respective density values are output to a spreadsheet. Quantify several equally sized, unstained areas of the gel to use as a background correction factor.

11. Average the density values for the unstained, background areas of the gel. Subtract the average background density from each of the density values generated for the BSA standard curve and the density values for purified component X.

12. To generate the standard curve, plot the density values for each of the BSA standards against the relative concentration that was loaded onto the gel. Use linear regression to draw a line through the points generated by the BSA standards and generate an equation of the line (Fig. 2b).

13. Substitute each density value that was generated for each amount of Hh component X that was loaded onto the gel (e.g., density values corresponding to duplicate volumes of 5, 10, and 15 µl of sample) into the standard curve equation as "y" and solve for "x" to estimate the relative concentration of the sample in each lane (Fig. 2b). Divide by the number of µl loaded for each sample to get an estimate of ng/µl in the Hh component X sample. Average the Hh component X values generated from the different volumes to get an averaged estimate of the sample's concentration.

14. Repeat **steps 9–13** to quantify proteins on stained gel 2.

15. If the density values of the purified Hh component protein X do not fall within the range of the BSA standard curve, you must adjust the amount of purified Hh component X or the range of the BSA standard curve and repeat steps above to rerun and stain gels (*see* **Note 2**).

16. Compare the concentration values generated by the two staining methods. As different staining methods can give variable results, we used two standard staining methods—Coomassie staining [10] and silver staining [11]—to quantify our purified protein levels and average the concentration estimates. If the results of the two methods differed substantially, we ran a third gel and stained the proteins with an additional protein staining technique, such as SYPRO Ruby per manufacturer's instructions to generate a third estimate of protein concentration and average the three estimates.

3.2 Quantify Amount of Endogenous Hh Component X in Cell/ Tissue Lysates Against a Standard Curve of Purified Hh Component X

1. Now that the concentration of the purified, recombinant Hh component X sample has been determined, it can be used as a standard to measure the concentration of the endogenous levels of Hh component X in a cell lysate (Fig. 3).

2. Prepare a lysate of cell line or tissue of interest. In our quantitation, we used Cl-8 cells that had been lysed in hypotonic lysis buffer, by douncing, and subjected to a low-speed ($2,000 \times g$) centrifugation step to remove the nuclei.

3. Quantify the concentration of the cell lysate, using a standard technique, such as BCA or Bradford protein assay.

4. Using lysis buffer as diluent, make four dilutions of the lysate with concentrations of 200 ng/µl, 400 ng/µl, 600 ng/µl, and 800 ng/µl. Prepare the samples for gel electrophoresis by diluting each lysate 1:1 with 2× Laemmli loading buffer and boiling

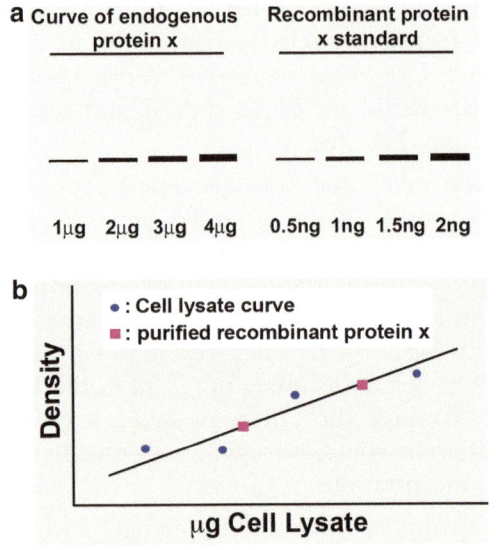

a Curve of endogenous   Recombinant protein
        protein x   x standard

1µg 2µg 3µg 4µg   0.5ng 1ng 1.5ng 2ng

b

• : Cell lysate curve
■ : purified recombinant protein x

Density

µg Cell Lysate

1. **Run purified recombinant Hh component protein X against cell lysates**

2. **Immunoblot to visualize endogenous Hh protein x in lysates and in recombinant Hh protein X standard**

3. **Quantify density values with image analysis software**

4. **Quantify endogenous Hh protein x in lysates against purified recombinant Hh component protein X values**

Fig. 3 Quantification of endogenous Hh component X in cell/tissue lysates against a standard curve of purified Hh component X. (**a**) Different volumes of cell lysate (containing endogenous Hh component X) and a standard curve of purified recombinant Hh component X are run on an SDS-PAGE gel and immunoblotted. (**b**) The densities of fluorescent immunoblotting signals are quantified, and the concentration of endogenous Hh component X is determined using the purified Hh component X standard curve

protein samples for 5 min. Final lysate dilution concentrations will be 100 ng/µl, 200 ng/µl, 300 ng/µl, and 400 ng/µl.

5. Prepare a gel for SDS-PAGE. When pouring the stacking gels, select combs with ~18 wells to ensure there are enough wells to accommodate the molecular weight standard, the purified component X standard curve, and eight lanes for the four dilutions of lysate in duplicate.

6. As the gels are polymerizing, prepare the purified component X samples for gel electrophoresis. Using the same buffer that purified X is suspended in as diluent, make four dilutions of the purified X of the following concentrations: 0.5 pg/µl, 1 pg/µl, 1.5 pg/µl, and 2 pg/µl. Prepare purified component X samples for gel electrophoresis by diluting each lysate 1:1 with 2× Laemmli loading buffer and boiling protein samples for 5 min.

7. Transfer the polymerized gel to the electrophoresis apparatus, fill the chambers with running buffer, and load the protein samples onto the gel with the protein samples (Fig. 3a). In the first lane, load a molecular weight standard. In lanes 2 through 9, load 10 µl of each cell lysate dilution in duplicate. When 10 µl of each dilution is loaded onto the gel, the final on-gel concentrations of the cell lysates will be 1 µg, 2 µg, 3 µg, and 4 µg. In lanes 10–17, load 4 µl of each purified X dilution in duplicate.

8. Run the gels until the molecular weight marker indicates that Hh component X is sufficiently resolved. Remove the gel from the electrophoresis chamber and transfer to a blotting membrane (PVDF or nitrocellulose) using standard transfer materials and equipment.

9. After transfer is completed, proceed with standard procedures to immunoblot for Hh component X, using primary antibodies to Hh component X and corresponding secondary detection antibodies (Fig. 3a).

10. When the blot is in the final postsecondary antibody washing steps, turn on fluorescent imaging equipment (e.g., STORM imaging system).

11. Remove the blot from the final wash and add ECL Plus immunoblotting substrate. Transfer the membrane to the imaging equipment and capture image according to manufacturer's instructions. Generally, a blue fluorescence setting will detect the ECL Plus signal.

12. Open the image of the blot with image analysis software, such as ImageQuant (GE Healthcare). Follow manufacturer's guidelines to quantify the density of each protein band individually, as previously done for gel images. Quantify several equally sized, blank areas of the membrane to use as a background correction factor (Fig. 3a).

13. Average the density values for the unstained, background areas of the blot. Subtract the average background density from each of the density values generated for the purified component X and the density values for endogenous component X.

14. To generate the standard curve, plot the density values for each of the purified component X standards against the relative concentration that was loaded onto the gel. Use linear regression to draw a line through the points generated by the purified component X standards and generate an equation of the line (Fig. 3b).

15. Substitute each density value that was generated for each amount of endogenous Hh component X that was detected in the lysate into the equation as "y" and solve for "x" to estimate the relative concentration of endogenous Hh component X the amount of lysate in each lane. Divide by the number of μl loaded on to the gel for each sample to get an estimate of pg Hh component X per ng of lysate. Average the Hh component X protein values generated from the different volumes to get an averaged estimate of the endogenous Hh component X concentration in lysate (Fig. 3b).

16. If the density values of the endogenous Hh component X do not fall within the range of the purified Hh component X curve, you must adjust the concentration of the lysate loaded

on the gel or the amount of purified Hh component X so both are within the same range and repeat steps above to rerun and quantify the protein levels (*see* **Note 3**).

17. Repeat the steps in Subheading 3.2 a minimum of three times to obtain a more accurate assessment of endogenous Hh component X levels.

4 Notes

1. Although most types of tagged, purified proteins may serve as a quantification standard, in our protocols we used the following:

   (a) FLAG-Fu: full-length Fu (amino acids (aa) 1–805) inserted into a baculoviral expression vector, transfected and grown in SF21 cells, and then purified as previously described [4–6].

   (b) FLAG-Cos2: full-length Cos2 (aa 1–1201) inserted into a baculoviral expression vector, transfected and grown in SF21 cells, and then purified as previously described [6].

   (c) FLAG-Ci: full-length Ci (aa 1–1398) inserted into a baculoviral expression vector, transfected and grown in SF21 cells, and then purified as previously described [6].

   (d) 6× His-Sufu: pRSET-Sufu was generated by inserting the full-length Sufu cDNA into the pRSET expression vector (Invitrogen), in frame with an internal 5′ 6×-histidine epitope tag. BL21(DE3)pLysS bacteria (Protein Express, Inc.) were transformed with the pRSET-Sufu, and 6× His-Sufu was purified under denaturing conditions using Ni NTA agarose beads (QIAGEN) per manufacturer's instructions.

   (e) A 6×-histidine-tagged amino terminal Smo peptide (amino acids 48–245) was generated and purified as previously described [7]. We chose to use this amino terminal fragment of Smo for its ease of purification and handling, as in our hands recombinant Smo tends to aggregate, making analysis and subsequent purification difficult. To ensure equal transfer to nitrocellulose membranes, we stained SDS-PAGE gels post-transfer to assess transfer efficiency of Smo peptide versus endogenous Smo. Calculations to determine the concentration of this smaller recombinant Smo peptide were converted from nanograms to moles in order to account for the difference in molecular weight between this peptide and endogenous Smo. Moreover, in order to compare these two proteins of different molecular weights, we also had to account for differences in their transfer efficiency during immunoblotting. We examined the difference in transfer out of a gel onto a nitrocellulose membrane, as well as the amount of protein that is able to transfer through the pores of a nitrocellulose membrane, for high and low molecular weight proteins. We found that the higher molecular weight, endogenous Smo transfers out of a gel 30 % less efficiently than the lower molecular weight Smo peptide. We also found that approximately 30 % of the small Smo peptide is lost through the pores of a nitrocellulose membrane. These transfer efficiency differences were factored into our calculations although the net change was negligible. Details of our calculations can be found in the supplemental experimental procedures section of Farzan et al. [6].

2. If the density values for the first set of volumes of Hh component X protein sample do not fall within the range of the BSA standard curve, try the following options: (1) adjust the volume of purified component X sample loaded on the gel by loading more or less, (2) adjust the volume of purified component X sample loaded on the gel by making a dilution of the stock, or (3) adjust the range of the BSA standard curve by loading more or less sample volume on the gel or making additional BSA dilutions (more dilute than 10 ng or less dilute than 200 ng). Be sure to record all changes to dilutions, standard curve, and volume loaded in order to calculate final concentrations. Keep in mind that protein staining techniques have upper and lower limits of detection, which vary by method.

3. Similarly, when blotting for endogenous protein levels, the endogenous and purified recombinant Hh component X levels must be within the same range for accurate quantification. Try the following options: (1) adjust the volume of purified component X sample loaded on the gel by loading more or less, (2) adjust the volume of purified component X sample loaded on the gel by making a dilution of the stock, or (3) adjust the endogenous level by loading more or less lysate volume on the gel or making additional lysate dilutions to expand range beyond 1–4 μg of lysate.

References

1. Farzan SF, Stegman MA, Ogden SK, Ascano M Jr, Black KE, Tacchelly O, Robbins DJ (2009) A quantification of pathway components supports a novel model of Hedgehog signal transduction. J Biol Chem 284:28874–28884

2. Lee E, Salic A, Kruger R, Heinrich R, Kirschner MW (2003) The roles of APC and Axin derived from experimental and theoretical analysis of the Wnt pathway. PLoS Biol 1:E10

3. Farzan SF, Ogden SK, Robbins DJ (2010) Quantitative insight into models of Hedgehog signal transduction. Fly (Austin) 4:141–144

4. Ascano M Jr, Nybakken KE, Sosinski J, Stegman MA, Robbins DJ (2002) The carboxyl-terminal domain of the protein kinase fused can function as a dominant inhibitor of hedgehog signaling. Mol Cell Biol 22:1555–1566

5. Ascano M Jr, Robbins DJ (2004) An intramolecular association between two domains of the protein kinase Fused is necessary for Hedgehog signaling. Mol Cell Biol 24:10397–10405

6. Stegman MA, Vallance JE, Elangovan G, Sosinski J, Cheng Y, Robbins DJ (2000) Identification of a tetrameric hedgehog signaling complex. J Biol Chem 275:21809–21812

7. Alcedo J, Zou Y, Noll M (2000) Posttranscriptional regulation of smoothened is part of a self-correcting mechanism in the hedgehog signaling system. Mol Cell 6: 457–465

8. Shaffer CD, Wuller JM, Elgin SC (1994) Raising large quantities of Drosophila for biochemical experiments. Methods Cell Biol 44:99–108

9. Stegman M, Robbins D (2007) Biochemical fractionation of Drosophila cells. Methods Mol Biol 397:203–213

10. Maniatis T, Fritsch EF, Sambrook J (1982) Molecular cloning: a laboratory manual. Cold Spring Harbor Laboratory, Cold Spring Harbor, NY

11. Blum H, Beier H, Gross HJ (1987) Improved Silver Staining of Plant-Proteins, RNA and DNA in Polyacrylamide Gels. Electrophoresis 8:93–99

Measuring Gli2 Phosphorylation by Selected Reaction Monitoring Mass Spectrometry

Robert Ahrends*, Pawel Niewiadomski*, Mary N. Teruel, and Rajat Rohatgi

Abstract

Phosphorylation is an important mechanism by which Gli proteins are regulated. When the Hedgehog (Hh) pathway is activated, multiple serine and threonine residues of Gli2 are dephosphorylated, while at least one residue undergoes phosphorylation. These changes in phosphorylation have functional relevance for the transcriptional activity of Gli proteins, as shown by in vitro and in vivo assays on Gli mutants lacking the phosphorylated residues. Here, we describe a method of quantitatively monitoring the phosphorylation of Gli proteins by triple quadrupole mass spectrometry of Gli2 immunoprecipitated from cell lysates. This method is broadly applicable to the monitoring of phosphorylation changes of immunoprecipitated Gli proteins when the putative phosphosites are known.

Key words Gli proteins, Phosphorylation, Protein kinase A, Triple quadrupole, Mass spectrometry, Selected reaction monitoring

Abbreviations

AmBic	Ammonium bicarbonate
ACN	Acetonitrile
FA	Formic acid
Hh	Hedgehog
HPLC	High-performance liquid chromatography
MS	Mass spectrometry
MS/MS	Tandem mass spectrometry
nLC	Nanoliter-scale high-performance liquid chromatography
SRM	Selected reaction monitoring

*R.A. and P.N. contributed equally and are listed alphabetically.

Natalia A. Riobo (ed.), *Hedgehog Signaling Protocols*, Methods in Molecular Biology, vol. 1322,
DOI 10.1007/978-1-4939-2772-2_10, © Springer Science+Business Media New York 2015

1   Introduction

Gli proteins, vertebrate homologs of the *Drosophila* protein Cubitus interruptus (Ci), are a family of transcription factors that mediate the effects of Hedgehog (Hh) signaling on gene expression [1]. The mammalian Gli2 and Gli3 proteins have long been known to be phosphorylated by protein kinase A (PKA) at six conserved serine/threonine residues, hereafter referred to as P1–P6, localized in their C-terminal part [2]. These phosphorylations were known to be required for the conversion of Gli2/Gli3 into truncated transcriptional repressors (GliR—Gli repressors) by the proteasome [3, 4]. More recently, we have shown that failure to phosphorylate these six residues results in the formation of constitutive Gli activators (GliA) capable of inducing the production of Hh target genes in the absence of any upstream signal [5]. Moreover, we showed that P1–P6 dephosphorylation is correlated with the phosphorylation of a serine at the N-terminus of Gli2, which we refer to as Pg [5]. Monitoring these phosphorylation events is important for the understanding of Gli protein regulation by PKA and other kinases.

Several methods can be used to measure protein phosphorylation in cells [6]. Labeling cells with radioactive [^{32}P]orthophosphate followed by electrophoresis (1-D or 2-D) and detection by film or phosphor imaging autoradiography is a well-established method in the field, but is generally limited to detection of phosphorylation at the level of a protein rather than quantification of phosphorylation at specific residues. Phosphospecific antibodies are an excellent tool for quantification of specific phosphosites [7]. The distinct advantage of this method is its sensitivity and ability to detect phosphoproteins by immunohistochemical methods in tissues and cells. However, a separate antibody must be produced for each phosphosite, which effectively limits the number of sites that can be monitored. In addition, phosphospecific antibodies of sufficient quality to provide quantitative information on phosphosite occupancy are often difficult to make, as has been the case with the Gli proteins.

Progress in mass spectrometric techniques has opened up new avenues for phosphosite analysis. In particular, large-scale "shotgun" LC-MS/MS-based screens have made it possible to discover and quantify new phosphosites in a global, high-throughput manner [8–10]. Chromatographic techniques based on metal ion affinity chromatography (IMAC) or titanium oxide are used to enrich phosphorylated species, thus reducing non-phosphorylated peptide abundance in the sample and increasing phosphopeptide detection sensitivity [10, 11]. However, these methods are unsuitable for the detection of less abundant phosphopeptides. In order to precisely quantify low-abundance phosphorylated peptide species, a targeted MS-based approach known as selected reaction monitoring (SRM) is often employed using a triple quadrupole instrument [12–14].

In the SRM experiment, proteins are digested by trypsin into short (5–50 amino acids) peptides. These peptides are resolved using reverse-phase high-performance liquid chromatography (HPLC) and ionized by electrospray ionization (ESI). After the peptide ions enter the instrument, the first quadrupole is used to select the desired precursor peptide ion that contains the phosphosite of interest. The precursor ion is then fragmented by collision-induced dissociation (CID) in the second quadrupole. The third quadrupole is used to select fragment ions that had been found to give the most intense signal in preliminary experiments, and these fragment ions reach the detector (Fig. 1). The combination of a targeted peptide precursor ion with one of its specific fragments is known as a "transition." Usually 2–3 transitions are enough to quantify the amount of a peptide in the extract using an internal standard. Since no survey mass spectra are acquired in the quantification mode, many peptides in the extract can be quantified relatively quickly (approximately 200 peptides in 1 h). The two-level mass filtering in the first and third quadrupole and the high ion transmission used in SRM account for a much higher sensitivity compared to conventional proteomic techniques [15]. These SRM approaches have been shown to detect specific peptides and phosphopeptides in complex mixtures [13].

The disadvantages of this method are that (1) only a limited number of fragment ions can be detected in each sample and (2) the precision of mass measurement is limited to around 0.3 Da around the precursor, sometimes leading to false positives with similar MS/MS fragmentation patterns being detected in place of the less abundant true peptide of interest. With those limitations in

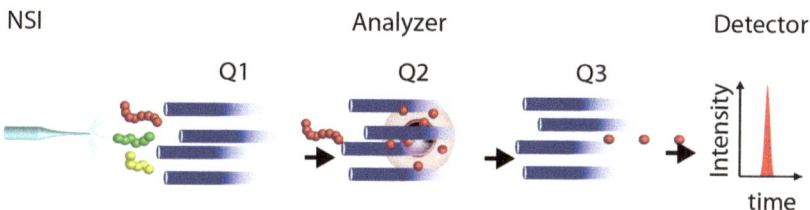

Fig. 1 Diagram illustrating the operation of the triple quadrupole instrument. The peptide solution is ionized in the nanospray ionization source (NSI). Precursor ions enter the first quadrupole (Q1). Precursor peptide ions within a narrow mass to charge (m/a) ratio (ideally representing a single precursor) pass through Q1 and enter Q2. In Q2, precursor ions are dissociated into fragment ions. Fragment ions enter Q3 and are filtered so that only fragment ions within a narrow m/a range pass onto the detector. A combination of a precursor ion and a specific fragment ion is known as a transition. The instrument cycles through multiple transitions and measures the intensity of each transition over time, which is known as the extracted ion current (XIC) for this transition (graph on the right). The total amount of the peptide in the solution is proportional to the area under the curve of the XIC

mind, the assay design must be undertaken with care to maximize sensitivity and specificity, and internal isotope-coded standard peptides have to be applied as an additional control.

Phosphopeptides of interest can be quantified and validated using isotope-labeled reference peptides in combination with SRM-MS. This simple method, based on the classical principle of isotope dilution, was introduced by the Gygi Lab in 2003 and uses synthesized isotope-labeled reference peptides as internal standards [13–16]. To develop a suitable reference peptide, a peptide out of the digested sample is selected based on the applied proteases, retention behavior, ionization efficiency, and fragmentation pattern. The selected peptide is then synthesized by solid-phase synthesis with light amino acids and one isotopically labeled "heavy" amino acid. This results in the synthesis of a chemically identical peptide homolog that differs from the endogenous peptide only in mass (8–10 Da). Since the endogenous (light) and the internal standard (heavy) peptides differ in mass, the heavy peptide transitions can be easily distinguished by triple quadrupole mass spectrometry from those of the endogenous peptide. Thus, the heavy peptide can be used as reference for the light peptide derived from the endogenous protein. Since the reverse-phase separation, ionization, and fragmentation are identical for the isotopically labeled and the corresponding endogenous peptides, the reference peptides elute at the same time as the endogenous peptides and are fragmented in an identical pattern. Consequently, if the concentration of the "heavy" peptide in the sample is known, the concentration of endogenous peptides can be directly measured as a ratio over the added isotopically labeled peptides (absolute quantification—AQUA). Alternatively, if the "heavy" peptide is added at an unknown but constant concentration to all samples, the relative increase or decrease in the abundance of the endogenous peptide can be quantified (relative quantification). The advantage of the latter method is that the isotopically labeled peptides do not need to be precisely titrated and can be purchased in an unpurified and therefore much more economical format.

Here, we describe relative quantification SRM assays that we designed for the measurement of the phosphorylation of Gli2 at five distinct sites, but the same methodology can be applied to any proteins that can be isolated at sufficient quantities and purity levels from cell lysates or tissues.

2 Materials

2.1 Cell Culture Materials

1. Cells: NIH/3T3 (ATCC).

2. High-serum culture media: DMEM containing 10 % fetal bovine serum (FBS), 1× GlutaMAX, 1× nonessential amino acids, 1× sodium pyruvate, 1× penicillin/streptomycin solution (Life Technologies).

3. Low-serum culture media: Same as high-serum media, except made with 0.5 % FBS instead of 10 %.

4. Cell culture dishes: Nunc Nunclon tissue culture-treated 10 and 15 cm diameter round dishes or equivalent.

5. Bortezomib stock solution: 10 mM bortezomib (LC Laboratories) in DMSO; store at −20 °C.

2.2 Cell Lysis/ Immunoprecipitation Materials

1. 1× phosphate-buffered saline (PBS): 137 mM NaCl, 2.7 mM KCl, 10 mM Na_2HPO_4, 1.8 m M KH_2PO_4; chill to 4 °C and adjust pH to 7.4.

2. Urea lysis buffer: 100 mM $NaH_2PO_4*H_2O$, 10 mM Tris base, 8 M urea, 300 mM NaCl, pH 8.0. Prepare fresh before each experiment and chill to 4 °C (*see* **Note 1**).

3. Cell lifters: Corning cell lifters or equivalent. Lifters can be rinsed in water, blotted dry, and reused.

4. High-salt RIPA buffer: 50 mM Tris, pH 7.4, 370 mM NaCl, 50 mM NH_4Cl, 25 mM glycinamide, 2 % Nonidet P-40 (or IGEPAL CA-630), 0.25 % sodium deoxycholate, 1 μM bortezomib, 1× SigmaFAST protease inhibitor cocktail, 1 mM β-glycero-phosphate, 1 μM microcystin LR, 1 mM NaF, 1 mM activated Na_3VO_4. Prepare fresh before each experiment and chill to 4 °C.

5. Ultracentrifuge and ultracentrifuge tubes that will hold at least 20 mL of liquid and withhold $100,000 \times g$ centrifugation.

6. Anti-Gli2 antibody: Any anti-Gli2 antibody suitable for immunoprecipitation can be used (*see* **Note 2**).

7. Anti-Gli2 beads: Couple anti-Gli2 antibody protein A-conjugated Dynabeads (Life Technologies). For each 100 μL of Dynabeads slurry, use 20–40 μg of the affinity-purified antibody. Cross-link antibody to beads using dimethylpimelidate. Wash thoroughly and store at 4 °C in 0.2 M ethanolamine, pH 8.5, 0.2 M NaCl, 0.1 % Nonidet P-40, 10 % glycerol, 0.01 % NaN_3.

8. Acid wash solution: 100 mM glycine, pH 2.5.

9. Magnetic holders for 1.5 mL microcentrifuge tubes and 15 mL or 50 mL conical tubes (DynaMag-2, DynaMag-15, DynaMag-50 from Life Technologies or equivalent).

10. Bead wash buffer: Mix 1 part urea lysis buffer with three parts high-salt RIPA buffer (final urea concentration 2 M).

11. SDS elution buffer: 50 mM Tris, pH 6.8, 2.5 % sodium dodecyl sulfate, 5 % glycerol. Add dithiothreitol to 65 mM before each experiment.

2.3 Gel Electrophoresis/ Staining Materials

1. Polyacrylamide gels: Novex Tris-Glycine 8 % mini-protein gels 1.5 mm, 10 wells (Life Technologies) or equivalent with an appropriate electrophoresis box (*see* **Note 3**).

2. Gel running buffer: 25 mM Tris base, 192 mM glycine, 0.1 % sodium dodecyl sulfate. Make a 10× concentrate and sterile-filter. Do not adjust pH. Dilute with ultrapure water before use.

3. Coomassie blue solution: GelCode Blue (Pierce) or equivalent Coomassie G-250 solution.

2.4 Gel Cleaning/ Digest/Extraction Materials

All reagents used at stages following gel fragment excision *must be HPLC or MS grade* to ensure the absence of trace contaminants that can interfere with nLC separation and MS detection.

1. Ammonium bicarbonate (AmBic) solution: 50 mM AmBic in HPLC-grade water. Store for a few weeks at 4 °C.

2. Acetonitrile (ACN): HPLC-grade acetonitrile (100 %).

3. AmBic/ACN solution: Mix 1 part AmBic solution with 1 part ACN. Store for a few weeks at 4 °C.

4. Vacuum centrifuge resistant to small quantities of organic solvents and a vacuum pump.

5. Trypsin solution: Sequencing-grade modified trypsin (Promega) resuspended in the supplied buffer to 200 ng/μL. Aliquot and store frozen at −80 °C.

6. Eppendorf Protein LoBind microcentrifuge 1.5 mL tubes.

7. ACN/formic acid (FA) solution: 5 % formic acid, 65 % ACN, 30 % HPLC-grade water.

8. FA solution: 5 % formic acid in HPLC-grade water.

9. 2 % ACN/0.1 % FA solution in HPLC-grade water.

10. Sonicating water bath.

11. Tabletop centrifuge with adapters for 96-well plates. We used the Beckman Allegra 6R with a GH-3.8A rotor.

12. Peptide cleanup plate: Waters Oasis HLB μElution 96-well plate (30 μm particle size).

13. Concentrated formic acid solution: 98 % formic acid.

14. 100 % methanol.

15. 0.04 % trifluoroacetic acid.

16. 80 % ACN solution in HPLC-grade water.

2.5 HPLC/MS Materials

1. Standard peptide mix: Heavy arginine ($[^{13}C]_6[^{15}N]_4$—total mass increase 10 Da)-labeled unphosphorylated and phosphorylated peptides can be purchased from multiple vendors. We have purchased peptides from JPT Peptide Technologies (SpikeTides-L) in the unpurified format. Solubilize peptides to approximately 1 mM in 80 % AmBic solution/20 % ACN and sonicate for 5 min in an ice-cold sonicating water bath. Add all the heavy peptides needed for the experiment to AmBic solution to a final concentration of 1 μM and store the peptide mix, as

well as the remainder of the concentrated individual peptide solutions, aliquoted at $-80\,^{\circ}$C (*see* **Note 4**).

2. HPLC buffers: A, 0.1 % FA in HPLC-grade water; B, 0.1%FA in ACN.

3. Nano-HPLC instrument: EASY-nLC (Proxeon) equipped with 35 mm $\times$ 0.1 mm C18 trapping column (Michrom C18, 5 μm, 120 Å) and a 200 mm $\times$ 0.075 mm diameter reverse-phase C18 capillary column (Maisch C18, 3 μm, 120 Å).

4. MS instrument: TSQ Vantage (Thermo Fisher Scientific) triple quadrupole instrument with a Proxeon nanospray ionization source.

2.6 Data Analysis Materials

1. Xcalibur 2.2.44, data analysis and instrument control.

2. Skyline 2.5 software suite—available for free from the MacCoss Lab website: https://brendanx-uw1.gs.washington.edu/labkey/project/home/software/Skyline/begin.view

3. MS Excel or any other statistical software.

3 Methods

3.1 Cell Culture

1. Culture NIH/3T3 cells in high-serum culture media (*see* **Note 5**). For the experiment, prepare 4–6 15 cm round dishes of fully confluent cells for each experimental condition (*see* **Note 6**).

2. After the cells have reached full confluence (look "over-crowded"), change the media to low-serum culture media for 36 h (*see* **Note 7**).

3. 4–6 h before harvesting, add bortezomib to 1 μM to block proteasomal degradation of Gli2.

4. Add hedgehog-modulating drugs as needed for 1–4 h before harvesting (*see* **Notes 8** and **9**).

3.2 Cell Lysis

All procedures in this section should be carried out at 4 $^{\circ}$C (in the cold room). Dishes can be processed in groups of 4–6.

1. Decant media from cells and wash briefly twice with ice-cold PBS.

2. Carefully remove as much of PBS as possible by aspirating or blotting the sides of each plate with a paper towel.

3. To each plate, add 0.8 mL of urea lysis buffer and scrape the cells off with a cell lifter. The cells will lyse as they are lifted, and the solution will become viscous. Pipet the solution into a 15 mL conical tube using a 1 mL tip with the end cut off.

4. Allow the cells to lyse for 15 min at 4 $^{\circ}$C (do not put on ice or the urea will precipitate).

5. Sonicate the lysate to reduce viscosity (*see* **Note 10**).

6. Dilute the lysate with high-salt RIPA buffer. Calculate the volume of the buffer to adjust the final urea concentration to 2 M.

7. Ultracentrifuge the samples at $100,000 \times g$ for 30 min to remove debris and protein aggregates. Keep a small aliquot of the supernatant as "total cell lysate."

8. Decant the supernatant into a conical tube.

3.3 Immunoprecipitation of Gli2

1. Prewash anti-Gli2 beads in acid wash solution for 5 min at room temperature. Wash twice with high-salt RIPA buffer. For each 15 cm dish of cells, use 16 µL of the bead slurry, corresponding to 3.2–6.4 µg of coupled antibody (*see* **Note 11**).

2. Add beads to lysate. Rotate at 4 °C for 24–40 h.

3. Magnetize beads on a magnetic stand. Keep a small aliquot of the supernatant as "flowthrough."

4. Remove supernatant from beads. Resuspend beads in 1 mL bead wash buffer and transfer to 1.5 mL microcentrifuge tubes.

5. Spin beads down briefly and magnetize. Aspirate the wash buffer and remove from the magnetic stand.

6. Resuspend beads in 50 µL SDS elution buffer.

7. Elute protein from beads with vigorous agitation at 37 °C for 30 min (*see* **Note 12**).

3.4 Electrophoresis and Gel Staining

From this point on, great care must be taken to avoid keratin contamination of samples. Always use fresh gloves, and if possible, perform all steps under laminar airflow or in a tissue culture hood. All containers and tools used for gel manipulation should be pre-cleaned by soaking overnight in 1 % SDS solution and rinsing several times with ultrapure water and kept clean by careful handling.

1. Load samples onto the gel and run the gels at 100 V constant voltage until the gel front has reached approximately 1/2 of the gel height. In order to avoid band "spreading" or "smiling," load SDS elution buffer into empty wells of the gel.

2. Remove the gel from its casing and wash 3×15 min in ultrapure water. Avoid touching the gel even with a gloved hand. Use a clean spatula instead.

3. Stain the gel with Coomassie blue according to the manufacturer's instructions. Destain the gel until clear bands at around 200 kDa are visible (*see* Fig. 2).

3.5 Gel Destaining and Trypsin Digestion

1. Transfer the gels onto a clean glass plate and excise the bands with a freshly opened scalpel blade. Cut the gel bands into pieces approximately 1 mm³ in size. Transfer gel pieces into sterile 1.5 mL microcentrifuge tubes (*see* **Note 13**).

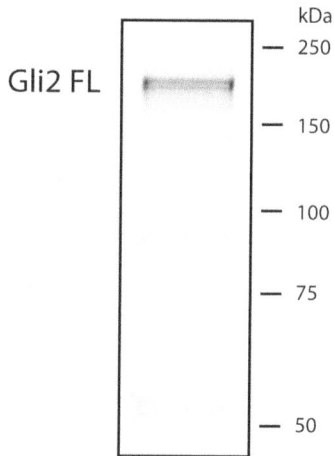

Fig. 2 Coomassie blue-stained gel of Gli2 immunoprecipitated from NIH/3T3 lysates

2. To each tube, add 250 μL AmBic solution. Mix vigorously at 37 °C for 30 min (a heating tube mixer, such as the Eppendorf ThermoMixer, can be used for this and subsequent steps). Spin down briefly.

3. Discard the supernatant and add 150 μL AmBic/ACN solution. Mix vigorously at 37 °C for 15 min. Spin down briefly.

4. Discard the supernatant, add 200 μL AmBic solution, and mix vigorously at 37 °C for 10 min. Spin down briefly.

5. Repeat **steps 3** and **4**. At this point, the gel pieces should be completely transparent without any traces of Coomassie blue staining. If there is still some blue staining remaining, **steps 3** and **4** can be repeated again.

6. Discard the supernatant. Add 200 μL ACN. Mix vigorously for 5 min. The gel slices should become completely opaque and dehydrated. Spin down briefly.

7. Discard the supernatant. Dry the gel pieces completely for 10 min in a vacuum centrifuge (*see* **Note 14**).

8. Put tubes with the dried gel pieces on ice. Take out one aliquot of trypsin solution and dilute 40× in ice-cold AmBic solution (final concentration, 5 ng/μL).

9. Add 30 μL of diluted trypsin to each tube with gel pieces. Leave on ice for 30 min until gel pieces swell and become transparent again.

10. Remove excess trypsin from gel pieces. Cover the pieces with 30 μL of AmBic solution.

11. Leave at 37 °C overnight (for 16–24 h).

3.6 Tryptic Digest Extraction

1. Add 10 µL of HPLC-grade water to each tube. Mix vigorously at 37 °C for 30 min. Spin down briefly.

2. Transfer supernatant to an Eppendorf Protein LoBind tube. Add 5 µL of the ACN/FA solution and 3 µL standard peptide mix to the tube with supernatant (*see* Table 1).

3. To the gel pieces, add 50 µL of FA solution. Mix vigorously at room temperature for 30 min. Spin down briefly.

4. Add the supernatant to the extract from **step 2**.

5. Repeat **steps 3** and **4**.

6. Repeat **step 3** and then sonicate gel pieces for 5 min in ice-cold water in a sonicating water bath. Spin down briefly.

7. Add supernatant to extracts from previous steps.

8. Repeat **steps 6** and 7, except use 100 µL of FA solution for the final extraction.

3.7 Peptide Cleanup (*See* Note 15)

1. Check the pH of the sample by placing 1–2 µL of the sample on a pH indicator paper. If the pH is above 3, acidify with 1–2 µL of concentrated formic acid.

2. Condition the columns of the peptide cleanup plate with 250 µl 100 % methanol.

4. Centrifuge at $200 \times g$ for 2 min. If liquid is still visible, repeat the centrifugation step.

5. Equilibrate the column with 250 µl of 0.04 % trifluoroacetic acid.

6. Centrifuge at $150 \times g$ for 2 min.

7. Load sample (up to 250 µl).

8. Centrifuge at $200 \times g$ for 3 min.

9. Wash sample with 250 µl of 0.04 % trifluoroacetic acid, pH 2–3.

10. Centrifuge at $150 \times g$ for 2 min.

11. Add 100 µl 80 % acetonitrile in H_2O to elute peptides.

12. Centrifuge at $150 \times g$ for 5 min.

13. Transfer each eluted peptide sample to a fresh Eppendorf Protein LoBind tube. Evaporate the peptide solution to complete dryness in a vacuum centrifuge (*see* **Note 16**). Resuspend the peptide pellet (which may not be visible at this point) in 15 µL of 2% ACN/0.1% FA solution. Pipet up and down 30 times and sonicate in an ice-cold sonicating water bath for 5 min.

14. Transfer the peptides into a nLC vial and store at 4 °C. If possible, analyze the sample on the same day.

Table 1
List of transitions for monitoring Gli2 phosphorylation

Phosphosite	Peptide sequence	Precursor neutral mass	Heavy precursor neutral mass	Precursor charge	Fragment ion	Fragment neutral mass	Heavy fragment neutral mass	Fragment charge
Control	AHTGGTLDDGIR	1211.6	1221.6	2	y9	902.4	912.5	1
					y8	845.4	855.4	1
					y5	574.3	584.3	1
Control	YAAATGGPPPTPLPGLDR	1749.9	1759.9	2	y11	1158.6	1168.6	1
					y10	1061.6	1071.6	1
					y7	766.4	776.4	1
P1	RDSS[+80]TSTM[+16]SSAYTVSR	1830.7	1840.8	3	y7	782.4	792.4	1
					y6	695.4	705.4	1
					y5	624.3	634.3	1
P2	RSS[+80]GISPYFSSR	1422.6	1432.6	2	y9	1012.5	1022.5	1
					y7	842.4	852.4	1
					y6	755.4	765.4	1
P5	RGS[+80]DGPTYSHGHGHGYAGAAPAFPHEGPNSSTR	3411.5	3421.5	5	y8	846.4	856.4	1
					y7	717.3	727.3	1
					b4–98	397.2	397.2	1
P6	RAS[+80]DPVRRPDPLILPR	1937.0	1947.0	3	y8	919.5	929.6	1
					y6	707.5	717.5	1
					b4–98	411.2	411.2	1
					b6–98	607.3	607.3	1

(continued)

Table 1
(continued)

Phosphosite	Peptide sequence	Precursor neutral mass	Heavy precursor neutral mass	Precursor charge	Fragment ion	Fragment neutral mass	Heavy fragment neutral mass	Fragment charge
dephospho-P6	ASDPVR	643,3	653,3	2	y5	572.3	582.3	1
					y4	485.3	495.3	1
					y3	370.2	380.2	1
Pg	TS[+80]PNSLVAYINNSR	1614.7	1624.7	2	y8	935.5	945.5	1
					y7	836.4	846.4	1
					y6	765.4	775.4	1
dephospho-Pg	TSPNSLVAYINNSR	1534.8	1544.8	2	y8	935.5	945.5	1
					y7	836.4	846.4	1
					y6	765.4	775.4	1

3.8 Selected Reaction Monitoring (SRM) Mass Spectrometry

1. Set up the following program for the separation/quantitation of the peptides:

   (a) Separation: linear gradient from 5 % to 45 % ACN over 70 min with a 300 nl/min flow rate for separation and a 4 μl/min flow rate for sample loading.

   (b) Positive polarity (tune file).

   (c) Resolution for Q1 and Q3: 0.7u FWHM (method file).

   (d) Emitter voltage: 1200–1,500 V (tune file).

   (e) Temperature of the transfer capillary: 270 °C (tune file).

   (f) Scheduled SRM: maximum window 5 min, cycle time 1 s, average dwell time 26 ms (*see* **Note 17**). Set the transitions as indicated in Table 1 (*see* **Note 18**).

2. Inject 3–5 μL of the sample and start the run. Repeat the run twice for each sample to account for run-to-run variability. Example extracted ion currents for one of our runs are shown in Fig. 3.

3. To verify proper operating performance and to determine if the LC/MS system needs cleaning or calibration, perform SRM on a standard peptide mixture (6 Bovine protein digest equal molar mix, Michrom, USA) at regular intervals between batches of samples.

3.9 Data Analysis

1. Import raw data into Skyline.

2. Export peak data from Skyline to a tab-separated text file (*see* **Note 19**). Export the following parameters: PeptideSequence, ProteinName, ReplicateName, PeptideRetentionTime, and RatioToStandard. Only export the transition with the highest peak for each sample and peptide.

3. Take the RatioToStandard as the "uncorrected abundance" of a specific peptide in the sample.

4. To correct for unequal loading and extraction of the sample from the gel, divide the uncorrected abundance of the non-phosphorylatable "control" peptides (first two entries in Table 1) by the average uncorrected abundance of the same peptide in the technical replicates of the untreated sample. Average of these values over the two control peptides will be denoted as the "loading ratio" of the sample. The loading ratio of the untreated sample is, by definition, 1.

5. Divide each uncorrected abundance value of the phosphorylated peptide of interest by the loading ratio for the respective sample to obtain the "corrected abundance value."

6. Divide the corrected abundance values by the average corrected abundance value of the untreated sample technical replicates to obtain fold increase of the specific phosphorylated or

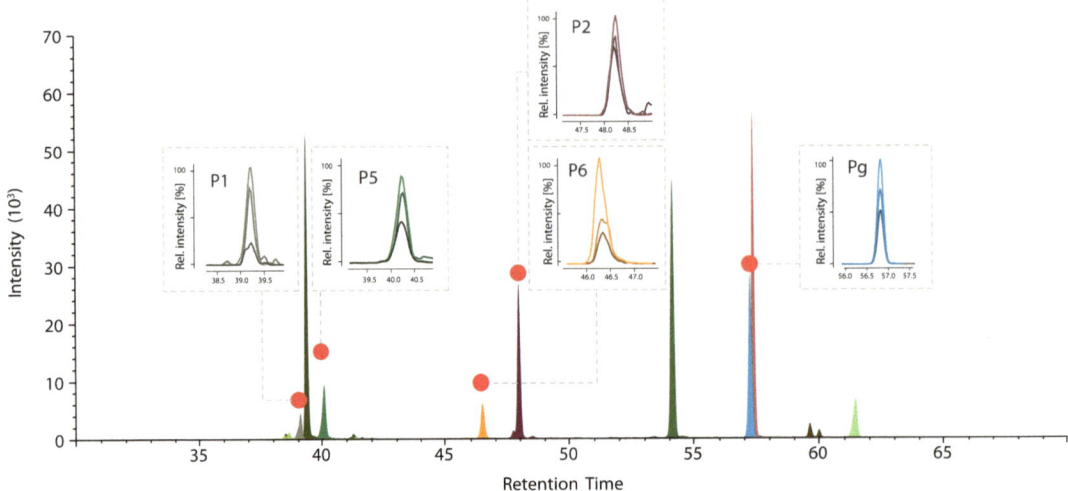

Fig. 3 Selected reaction monitoring (SRM) extracted ion currents (XICs) of Gli peptides. Chromatogram shows all monitored Gli phosphopeptides and Gli control peptides of approximately 250 ng Gli2 digest separated by reverse-phase nano-HPLC and analyzed by a triple quadrupole mass spectrometer. Each peptide XIC is displayed in a different color. The insets show XICs for individual transitions of the phosphopeptides used in this study

unphosphorylated peptide abundance over the untreated sample. The fold increase values should be reproducible between technical replicates and from experiment to experiment and can be reported. Example calculations are shown in Table 2.

4   Notes

1. The urea in the lysis buffer may precipitate in the cold. If it does, take the solution out to room temperature for a few minutes and resuspend the precipitated urea by vigorous mixing.

2. The anti-Gli2 antibody we used for immunoprecipitation of Gli2 was custom-made in rabbits from an antigen based on Cho et al. [17]. The antigen was a His-tagged fragment containing amino acids 1053–1264 of mouse Gli2 (VQYIKAH… AKPSHLG), and the sera were affinity purified on amino-link beads coupled to the same antigen.

3. We do not recommend pouring gels by hand, since it is very difficult to protect the gel from contamination with the ubiquitous keratins. There is no reason to use one brand over another, except the wells of the gel must be spacious enough to hold 50 μL of the sample.

4. If the peptides are designed for a new protein or a new phosphosite, they should only be ordered after an initial MS optimization phase to ensure identical sequences between the tryptic peptides from the sample and the synthetic standard peptides (*see* also **Note 17**).

Table 2
Example calculations of fold increase in the abundance of phosphopeptide P6 (RAS[+80]DPVRRPDPLILPR)

Site	Sample	Technical replicate	RatioToStandard = uncorrected abundance	Divided by control	Loading ratio	Corrected abundance	Fold increase
AHTGGTLDDGIR	Untreated	1	0.9035	1.006	1.000		
AHTGGTLDDGIR	SAG	1	0.8941	0.996	0.981		
AHTGGTLDDGIR	FSK	1	1.1101	1.236	1.277		
AHTGGTLDDGIR	Untreated	2	0.8923	0.994			
AHTGGTLDDGIR	SAG	2	0.8335	0.928			
AHTGGTLDDGIR	FSK	2	1.1501	1.281			
YAAATGGPPPTPLPGLDR	Untreated	1	0.4964	0.995			
YAAATGGPPPTPLPGLDR	SAG	1	0.4965	0.995			
YAAATGGPPPTPLPGLDR	FSK	1	0.6482	1.299			
YAAATGGPPPTPLPGLDR	Untreated	2	0.5015	1.005			
YAAATGGPPPTPLPGLDR	SAG	2	0.5016	1.005			
YAAATGGPPPTPLPGLDR	FSK	2	0.6436	1.290			
RAS[+80]DPVRRPDPLILPR	untreated	1	1.298			1.298	1.003
RAS[+80]DPVRRPDPLILPR	SAG	1	0.6155			0.627	0.485
RAS[+80]DPVRRPDPLILPR	FSK	1	3.5887			2.811	2.173
RAS[+80]DPVRRPDPLILPR	Untreated	2	1.2892			1.289	0.997
RAS[+80]DPVRRPDPLILPR	SAG	2	0.6886			0.702	0.543
RAS[+80]DPVRRPDPLILPR	FSK	2	3.8788			3.038	2.349

Non-phosphorylatable peptides AHTGGTLDDGIR and YAAATGGPPPTPLPGLDR are used for loading ratio calculations

5. Do not allow for the cells to reach full confluence until right before the experiment. It is best to split the cells every 2 days to keep them between 10 and 70 % confluent.

6. Full confluence is important for cell responsiveness to the signal and will also determine the quantity of Gli2 that can be isolated, which is decisive for the sensitivity of the method. For preliminary experiments, it may be worthwhile to increase the number of plates to 8–12 per condition.

7. Longer serum starvation seems to increase responsiveness, but will also make the cells more fragile. Starving for more than 40 h dramatically decreases cell viability.

8. For maximal dephosphorylation of sites P1–P6 and maximal phosphorylation of site Pg, 4 h treatment with 100 nM Smoothened agonist (SAG) is sufficient.

9. It is difficult to lyse more than 6 dishes at the same time, so treatment times must be spaced appropriately to allow time between batches of lysis.

10. Settings for the sonication must be adjusted for each sonicator individually. Avoid overheating the solution—use pulsed sonication in the cold room. The solution after the sonication should be easy to pipet with a standard 1 mL pipette tip.

11. Bead/antibody quantities should be adjusted based on the affinity of the antibody used. A pilot experiment with different bead to lysate ratios can be used for this purpose.

12. The elution from the antibody we used seems to be optimal at 37 °C, but with different antibodies, there may be some variability in the optimal elution conditions.

13. Gel pieces can be frozen at this point—add 50 μL of AmBic solution and place at –80 °C.

14. Dry gel pieces tend to be statically charged and may fly off the tubes through attraction to other statically charged objects, such as gloves.

15. We have found that peptide cleanup prior to running the reaction dramatically improved sensitivity. However, the cleanup may result in some losses of the material, so it is recommended to try different C18 materials for the cleanup in the optimization phase (Oasis HLB μElution 96-well plates worked best in our hands). One plate can be used for multiple experiments, but individual columns (wells) can be used only once. Mark columns that had been used to avoid contaminating your samples. Condition and equilibrate only the columns that will be used for the experiment in progress.

16. Drying the peptides may take a long time (up to 6 h), depending on how deep the vacuum is. If possible, use a high-performance solvent-resistant vacuum centrifuge connected to a vacuum pump of sufficient power.

17. Scheduled SRM must first be optimized in an unscheduled SRM run using a standard peptide mixture. Retention times for specific transitions may vary slightly from run to run depending on the composition of the sample and the performance of the instrument. Using scheduled rather than unscheduled SRM for the actual sample runs can dramatically improve sensitivity, since fewer transitions are monitored at the same time.

18. Selection of the right transitions for the phosphosites of interest takes significant up-front effort. For a new protein or a new phosphorylation site, it is recommended to initially run the sample on a tandem MS instrument to obtain an unbiased view of the peptide coverage of the protein, the tryptic digest pattern, and fragmentation spectra of specific peptides. Trypsin cleavage can be affected by phosphorylation, so the most abundant precursor peptides may have different lengths for the unphosphorylated and the phosphorylated peptide (see the P6 peptide of Table 1). For sites where phosphorylation stoichiometry is predicted to be significant (>20 %), it is sometimes advisable to monitor both the phosphorylated and the unphosphorylated peptide, so that absolute phosphorylation stoichiometry can be calculated. Even though fragmentation patterns may differ slightly between the tandem MS instrument and the triple quadrupole instrument, data obtained from a shotgun tandem MS experiment are a fairly good starting point for the setup of SRM transitions. Initially, at the optimization stage using a triple quadrupole instrument, it is recommended to try 5–6 transitions for each peptide, including transitions resulting from neutral loss of the phosphate moiety (–98 Da). Later, 2–3 most intensive transitions per peptide are selected for the actual experiments in order to maximize the dwell time and increase sensitivity. Since the standard heavy-labeled peptides are fairly expensive, they should only be ordered once the best precursor peptides and transitions have been selected based on the sample of interest. In the second stage of optimization, the heavy standard peptide extracted ion currents (peak height vs. retention time) are compared to those of the putative light peptide in the sample. If the retention time or fragmentation pattern differs between the heavy and the light peptide, chances are that the peptide detected in the sample is a false positive. In such case, the design of the peptide transitions should be started from scratch.

    It is also important to consider additional phosphorylation sites within the peptide of interest. It is possible that the changes in phosphorylation that one observes can be attributed to one of those additional sites, rather than the one we are interested in. If the transition peak retention time and relative intensities are identical between the standard heavy and the light sample peptide, we can be fairly confident that the peptide we are monitoring is the same as the known standard peptide.

Oxidation status of methionine is another confounding factor for some peptides (see peptide P1 in Table 1). We were able to detect both the unoxidized and the oxidized form of P1, but the oxidized form was more abundant under standard handling conditions. We found that in vitro reduction of oxidated methionine resulted in some sensitivity loss, and therefore we used the oxidated peptide as our SRM precursor.

Finally, it must be mentioned that not all phosphorylation sites are amenable to SRM quantification. Some tryptic peptides are either too short or too long or have physicochemical properties that make them either elute poorly from the LC column or ionize/fragment poorly in the MS instrument. For those reasons, we were unable to quantify phosphorylation at a few sites that we were interested in (P3, P4).

19. Automatically assigned peak widths must sometimes be adjusted manually prior to export. Inspect automatic peaks for inaccuracies and artifacts. Peak boundaries must be the same for the sample peptide and the corresponding heavy standard peptide, but may differ slightly between runs.

Acknowledgements

This work was supported by the OPUS grant from the Polish National Science Centre (grant 2014/13/B/NZ3/00909) to P.N., the following grants from the US National Institutes of Health: R21NS074091 to R.R., P50GM107615 to M.N.T. and 1R01DK10174301 to M.N.T., a Distinguished Scientist Award from the Sontag Foundation to R.R., a Scholar award from the Pew Foundation to R.R., a Scholar award from the V Foundation for Cancer Research to R. R., Stanford BioX to M.N.T., and the German Research Foundation grant AH 220/1-1 to R.A..

References

1. Hui C-C, Angers S (2011) Gli proteins in development and disease. Annu Rev Cell Dev Biol 27:513–537

2. Wang B, Fallon JF, Beachy PA (2000) Hedgehog-regulated processing of Gli3 produces an anterior/posterior repressor gradient in the developing vertebrate limb. Cell 100:423–434

3. Wang G, Wang B, Jiang J (1999) Protein kinase A antagonizes Hedgehog signaling by regulating both the activator and repressor forms of Cubitus interruptus. Genes Dev 13:2828–2837

4. Pan Y, Wang C, Wang B (2009) Phosphorylation of Gli2 by protein kinase A is required for Gli2 processing and degradation and the Sonic Hedgehog-regulated mouse development. Dev Biol 326:177–189

5. Niewiadomski P, Kong JH, Ahrends R, Ma Y, Humke EW, Khan S et al (2014) Gli protein activity is controlled by multisite phosphorylation in vertebrate Hedgehog signaling. Cell Rep 6:168–181

6. De Graauw M, Hensbergen P, van de Water B (2006) Phospho-proteomic analysis of cellular signaling. Electrophoresis 27:2676–2686

7. Blaydes JP, Vojtesek B, Bloomberg GB, Hupp TR (2000) The development and use of

phospho-specific antibodies to study protein phosphorylation. Methods Mol Biol 99: 177–189

8. Yi T, Zhai B, Yu Y, Kiyotsugu Y, Raschle T, Etzkorn M et al (2014) Quantitative phosphoproteomic analysis reveals system-wide signaling pathways downstream of SDF-1/CXCR4 in breast cancer stem cells. Proc Natl Acad Sci U S A 111:E2182–E2190

9. Beck F, Geiger J, Gambaryan S, Veit J, Vaudel M, Nollau P et al (2014) Time-resolved characterization of cAMP/PKA-dependent signaling reveals that platelet inhibition is a concerted process involving multiple signaling pathways. Blood 123:e1–e10

10. Engholm-Keller K, Larsen MR (2013) Technologies and challenges in large-scale phosphoproteomics. Proteomics 13:910–931

11. Bodenmiller B, Mueller LN, Mueller M, Domon B, Aebersold R (2007) Reproducible isolation of distinct, overlapping segments of the phosphoproteome. Nat Methods 4: 231–237

12. Abell E, Ahrends R, Bandara S, Park BO, Teruel MN (2011) Parallel adaptive feedback enhances reliability of the Ca2+ signaling system. Proc Natl Acad Sci U S A 108: 14485–14490

13. Gerber SA, Rush J, Stemman O, Stemman O, Kirschner MW, Gygi SP (2003) Absolute quantification of proteins and phosphoproteins from cell lysates by tandem MS. Proc Natl Acad Sci U S A 100:6940–6945

14. Ahrends R, Ota A, Kovary KM, Kovary KM, Park BO, Teruel MN (2014) Controlling low rates of cell differentiation through noise and ultrahigh feedback. Science 344:1384–1389

15. Stahl-Zeng J, Lange V, Ossola R, Eckhardt K, Krek W, Aebersold R, Domon B (2007) High sensitivity detection of plasma proteins by multiple reaction monitoring of N-glycosites. Mol Cell Proteomics 6:1809–1817

16. Kirkpatrick DS, Gerber SA, Gygi SP (2005) The absolute quantification strategy: a general procedure for the quantification of proteins and posttranslational modifications. Methods 35:265–273

17. Cho A, Ko HW, Eggenschwiler JT (2008) FKBP8 cell-autonomously controls neural tube patterning through a Gli2- and Kif3a-dependent mechanism. Dev Biol 321:27–39

Chapter 11

Rapid Screening of Gli2/3 Mutants Using the Flp-In System

Pawel Niewiadomski and Rajat Rohatgi

Abstract

Gli2 and Gli3 respond to the Hedgehog (Hh) signal in mammals by undergoing posttranslational modifications and moving to the nucleus. The study of Gli proteins has been hampered by the fact that their overexpression in cells prevents their proper regulation. To address this issue, we have developed a method of rapid generation of stable cell lines expressing near-endogenous and approximately equal levels of wild-type and mutant Gli proteins. This method is applicable to the study of effects of various mutations on Gli protein modifications and activity.

Key words Gli proteins, Flp-In system, Stable transfection, Flp-mediated recombination

1 Introduction

The three mammalian Gli proteins translate the Hh signal into transcriptional output of the pathway in the nucleus [1]. Gli2 and Gli3 undergo posttranslational modifications and subcellular localization changes in response to upstream signals [2–5]. Only when these proteins are expressed at endogenous or near-endogenous levels is this physiological regulation preserved. If the levels of expression are too high, Gli2 and Gli3 escape proper regulation by the Hh pathway and activate gene expression in the absence of upstream signal [4]. For these reasons, the study of the effects of mutations on Gli2/3 activity has been challenging. The two options were to either generate transgenic or knock-in mice bearing the mutation in question [6] or test multiple clones of stably transfected cell lines to select the ones with the right amount of exogenous Gli2/3 expression [7]. Both of these strategies require significant effort and are not suitable for rapid screening of a large number of mutations. Recently, we have devised a strategy to streamline the production of stable cell lines expressing near-endogenous levels of Gli2 and Gli3 based on Flp-In cells [4].

Natalia A. Riobo (ed.), *Hedgehog Signaling Protocols*, Methods in Molecular Biology, vol. 1322,
DOI 10.1007/978-1-4939-2772-2_11, © Springer Science+Business Media New York 2015

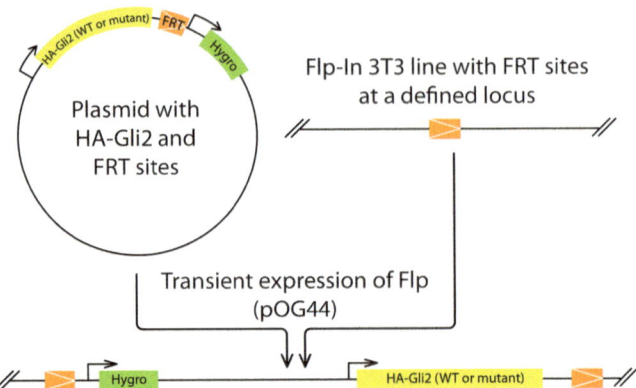

Fig. 1 Schematic representation of stable line generation using the Flp-In system. The Flp-In 3T3 cell line is co-transfected with a plasmid carrying the 3 × HA-tagged sequence of Gli2/3 (WT or mutant) and the pOG44 plasmid encoding Flp recombinase. Flp induces recombination between FRT site on the plasmid and a corresponding site in the genome of the Flp-In line. Stable line selection is accomplished by hygromycin

The Flp-In system, developed by Life Technologies, depends on Flp-mediated recombination between Flippase Recognition Target (FRT) sites [8] in the host cell and the transfected plasmid. It consists of three elements: Flp-In cell lines, the pOG44 Flp-recombinase expression vector, and a series of vectors, each containing an FRT site plus a hygromycin resistance gene (Fig. 1). Each Flp-In cell line has an FRT recombination site integrated into the genome (single copy per cell). When pOG44 is co-transfected with an FRT site-containing vector, the FRT plasmid is linearized and inserted into the defined genomic locus of the Flp-In line by the Flp recombinase. Subsequently, stable integrants are selected using hygromycin. Here, we describe technical details of the procedure in the specific case of stable lines that express near-endogenous levels of Gli2 and Gli3 mutants.

This approach can be used to gain insights into how mutations in Gli2 and Gli3 regulate their intrinsic transcriptional activity, subcellular localization, posttranscriptional modifications, and stability. However, it is important to note that Flp-In NIH/3T3 cells continue to express endogenous Gli2 and Gli3 proteins. Thus, if the goal is to analyze how mutations impact ligand-induced activation of Hh target genes, these wild-type endogenous proteins must be depleted, using either RNAi against the untranslated regions of the endogenous transcript or genetic deletion, for example, using the CRISPR/Cas9 system.

2 Materials

2.1 Cell Culture Materials

1. Cells: Flp-In 3T3 cells (Life Technologies).
2. Antibiotics-free media: DMEM containing 10 % fetal bovine serum (FBS), 1× GlutaMAX, 1× nonessential amino acids, 1× sodium pyruvate (all from Life Technologies).
3. High-serum culture media: antibiotics-free media with 1× penicillin/streptomycin solution (Life Technologies).
4. 0.05 % trypsin-EDTA solution (Life Technologies or equivalent).
5. Sterile phosphate-buffered saline (PBS), calcium- and magnesium-free.
6. Cell culture dishes: Nunc Nunclon or equivalent tissue culture-treated 6 cm and 10 cm round dishes and well plates (12- and 6-well).
7. Cell counter.

2.2 Materials for Stable Transfection and Cell Line Testing

1. FRT plasmid: plasmid containing an FRT site, hygromycin resistance gene, and the coding sequence for Gli2 or Gli3 tagged with the triple hemagglutinin (3×HA) tag on the N-terminus (*see* **Note 1**), stored at 100 ng/μL in water or TE buffer.
2. pOG44 plasmid stored at 1 μg/μL in water or TE buffer.
3. Cell transfection reagent: jetPRIME (Polyplus) or equivalent (*see* **Note 2**).
4. Hygromycin selection media: high-serum culture media containing 200 μg/mL hygromycin (Life Technologies or equivalent).
5. SDS-PAGE electrophoresis and western blotting equipment and materials.
6. Anti-HA antibody (HA-11 clone 16b12, Covance cat. # MMS-101P, used at 1:1,000).

3 Methods

3.1 Cell Culture

1. Culture Flp-In-3T3 cells in high-serum culture media. Split cell every 2–3 days to prevent cells from reaching over 80 % confluence.
2. To passage cells, wash once with PBS and cover with pre-warmed trypsin-EDTA solution for 3–5 min at 37 °C. Tap to dislodge cells and neutralize trypsin with high-serum culture media (for normal passage) or antibiotics-free media (before transfections). Count cells and plate at desired density.

3.2 Stable Transfection

1. Plate 5×10^5 cells on a 6 cm culture dish in 4 mL of antibiotics-free media. Immediately proceed to preparation of transfection mix (day 1, *see* **Note 3**).

2. Prepare transfection mix as recommended by the manufacturer of the transfection reagent. For each dish, use 300 ng of the FRT plasmid and 2.7 μg of the pOG44 plasmid.

3. Transfect cells as recommended by the transfection reagent manufacturer.

4. Leave cells overnight in a cell culture incubator.

5. On day 2, change media to fresh antibiotics-free media.

6. On day 3, rinse cells with PBS and trypsinize. Neutralize trypsin with high-serum culture media and split cells from each 6 cm dish into 3×10 cm dishes.

7. On day 4, change media to hygromycin selection media.

8. On subsequent days, split cells as needed to prevent overgrowth (keep cells below 80 % confluence; *see* **Note 4**).

9. Change media every 2–3 days to fresh hygromycin selection media.

10. Once most nonresistant cells die (after approximately 1 week), search for resistant cell clones. When the clones have grown to approximately 20–50 cells each, trypsinize cells and pool them. Depending on how many clones there are, plate cells into a single well in a 12-well or a 6-well plate in hygromycin selection media (*see* **Note 5**).

11. On subsequent days, split cells as needed. Cells no longer need to be maintained in hygromycin selection media all the time, but using hygromycin every 2–3 passages helps prevent downregulation of HA-Gli2/Gli3 expression (*see* **Note 6**).

12. Test expression and activity of HA-tagged Gli2 or Gli3 by anti-HA and anti-Gli1 immunoblot. *See* Fig. 2 for sample blots.

4   Notes

1. We cloned $3 \times$ HA-tagged Gli2 and Gli3 and their mutants into the pENTR2B vector and used Gateway cloning to insert the construct into the pEF5/FRT/V5-DEST. The vectors and Gateway cloning kit were purchased from Life Technologies. In principle, different FRT-carrying vectors should also work but may result in varying expression levels. Ready-made pEF5/FRT/V5-DEST plasmids containing $3 \times$ HA-tagged Gli2 and Gli3 WT and various mutant variants were deposited by us in the Addgene repository (*see* https://www.addgene.org/Rajat_Rohatgi/).

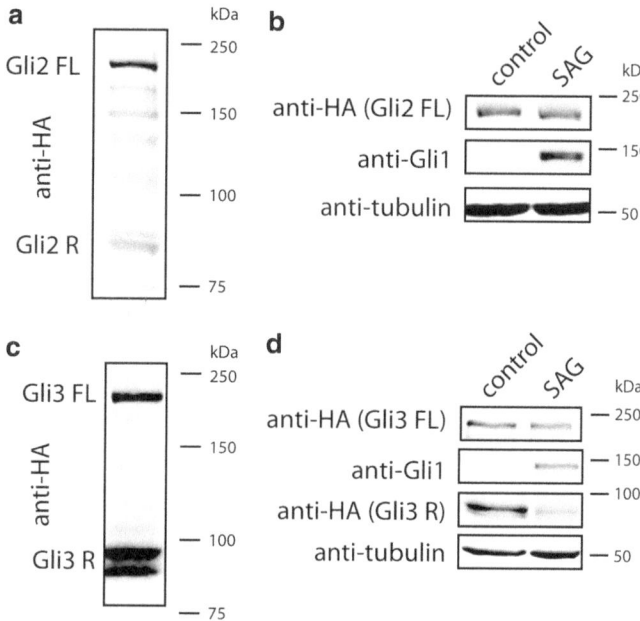

Fig. 2 Sample immunoblots of lysates from stable lines expressing wild-type 3×HA-Gli2 (**a**, **b**) or 3×HA-Gli3 (**c**, **d**) at near-endogenous levels. (**a**) Anti-HA immunoblotting of lysate from the cell line expressing WT 3×HA-Gli2. FL indicates the full-length form of the protein; R indicates the proteolytically processed repressor form. (**b**) Cells expressing 3×HA-Gli2 were left untreated or treated with 100 nM smoothened agonist (SAG) for 24 h. Full-length 3×HA-Gli2 was detected with anti-HA antibodies. Also shown are levels of Gli1 measured by anti-Gli1 immunoblotting. Tubulin is used as loading control. (**c**) Anti-HA blot of lysate from the cell line expressing WT 3×HA-Gli3. (**d**) Cells expressing 3×HA-Gli3 were treated as in (**b**). Full-length and repressor forms of 3×HA-Gli3 were detected with anti-HA antibodies. Note lack of Gli1 expression in control samples in (**b**) and (**d**), which indicates that the exogenous Gli2 and Gli3 proteins are properly regulated and do not induce pathway activation in the absence of treatment

2. We found the jetPRIME reagent to be most efficient for transfection of NIH/3T3-based cell lines, but other reagents, such as the X-tremeGENE 9 (Roche), also work reasonably well.

3. Best transfection efficiencies were achieved in our hands when the cells were transfected up to 2 h after plating, instead of 12–16 h after plating as recommended by most transfection reagent manufacturers.

4. Generally, hygromycin takes about a week to start killing non-resistant cells off. Once most cells are dead, there will be few (up to ten) clones in each culture dish, so discarding cells is not recommended. If possible, split cells into more dishes rather than throwing the extra cells away.

5. Alternatively, single clones may be picked and expanded separately, rather than pooled. Although the site of integration of the transgene is always the same in the Flp-In system, individual clones may have slightly different expression levels and/or growth characteristics, which may affect downstream assay results.

6. Stable cell lines generated using the Flp-In method tend to silence the exogenous Gli proteins over a large number of passages. It is recommended to cryopreserve early passages of stable lines and to not use these lines for longer than 15–20 passages.

Acknowledgments

This work was supported by the OPUS grant from the Polish National Science Centre (grant 2014/13/B/NZ3/00909) to P.N., a grant from the US National Institutes of Health (R21 NS074091) to R.R., a Distinguished Scientist Award from the Sontag Foundation to R.R., a Scholar award from the Pew Foundation to R.R., and a Scholar award from the V Foundation for Cancer Research to R.R.

References

1. Hui C-C, Angers S (2011) Gli proteins in development and disease. Annu Rev Cell Dev Biol 27:513–537

2. Humke EW, Dorn KV, Milenkovic L, Scott MP, Rohatgi R (2010) The output of Hedgehog signaling is controlled by the dynamic association between Suppressor of Fused and the Gli proteins. Genes Dev 24:670–682

3. Tukachinsky H, Lopez LV, Salic A (2010) A mechanism for vertebrate Hedgehog signaling: recruitment to cilia and dissociation of SuFu-Gli protein complexes. J Cell Biol 191:415–428

4. Niewiadomski P, Kong JH, Ahrends R, Ma Y, Humke EW, Khan S et al (2014) Gli protein activity is controlled by multisite phosphorylation in vertebrate Hedgehog signaling. Cell Rep 6:168–181

5. Niewiadomski P, Zhujiang A, Youssef M, Waschek JA (2013) Interaction of PACAP with Sonic hedgehog reveals complex regulation of the hedgehog pathway by PKA. Cell Signal 25:2222–2230

6. Pan Y, Wang C, Wang B (2009) Phosphorylation of Gli2 by protein kinase A is required for Gli2 processing and degradation and the Sonic Hedgehog-regulated mouse development. Dev Biol 326:177–189

7. Kim J, Kato M, Beachy PA (2009) Gli2 trafficking links Hedgehog-dependent activation of Smoothened in the primary cilium to transcriptional activation in the nucleus. Proc Natl Acad Sci U S A 106:21666–21671

8. McLeod M, Craft S, Broach JR (1986) Identification of the crossover site during FLP-mediated recombination in the Saccharomyces cerevisiae plasmid 2 microns circle. Mol Cell Biol 6:3357–3367

Chapter 12

Insights into Gli Factors Ubiquitylation Methods

Paola Infante, Romina Alfonsi, and Lucia Di Marcotullio

Abstract

The Hedgehog (Hh) signaling pathway governs cell growth and tissue development. Malfunctioning of several Hh pathway components, including the key transcriptional effector Gli proteins, is responsible for the onset of several tumors. Gli proteins activity is finely controlled by multilayered regulatory mechanisms, the most prominent of which is their proteasome-dependent proteolytic cleavage or massive ubiquitin-mediated proteolysis. Here, we described multiple procedures to determine whether a Gli protein is ubiquitylated both in a cellular context and in vitro, in basal conditions or by different E3 ubiquitin ligases and whether these processes are associated to Gli proteasome degradation.

Key words Gli factors, Ubiquitylation in vivo, Ubiquitylation in vitro, Recombinant proteins, Western blot

---

1 Introduction

Ubiquitylation is a posttranslational modification that regulates key cellular pathways [1], including the Hedgehog (Hh) signaling, a conserved developmental pathway crucial for tissue patterning, stem cell maintenance, and tumorigenesis [2]. The effects of this signaling are mediated by the Gli family of transcription factors (Gli1, Gli2, and Gli3), which act on a set of target genes promoting cell proliferation and preventing cell differentiation. Gli2 and Gli3 can be processed into a truncated N-terminal repressor form, whereas Gli1 behaves exclusively as a strong transcriptional activator [3]. Modulation of Gli activity occurs at multiple levels, including several regulators and posttranslational modifications. Ubiquitin-dependent processes have emerged as critical mechanisms by which Gli proteins stability, activity, or localization is controlled [2, 4, 5].

Ubiquitylation involves a cascade of reactions of different enzymes (E1, E2, E3) and terminates with the transfer of ubiquitin to substrate proteins as polyubiquitin chains or by monoubiquitylation or multi-monoubiquitylation events [1, 6]. Each one of

Natalia A. Riobo (ed.), *Hedgehog Signaling Protocols*, Methods in Molecular Biology, vol. 1322,
DOI 10.1007/978-1-4939-2772-2_12, © Springer Science+Business Media New York 2015

these poly-, mono-, or multi-monoubiquitylation events results in a different fate of ubiquitylated target protein and functional outcome [1, 7]. Protein polyubiquitylation frequently leads to total proteolytic processing of the substrate, allowing recognition by the 26S proteasome system, but can also induce the cleavage of the target protein and partial degradation. In this series of enzymatic reactions, the major determinants are the E3 ubiquitin ligases that provide specificity for substrate recognition [8].

Gli ubiquitylation is controlled by a number of E3 ligases belonging to the RING–Cullin, such as Cullin1-Slimb/βTrCP- and Cullin3-HIB/Roadkill/SPOP [9–11], and HECT family (Homologous to E6-AP Carboxy Terminus), such as Itch [12, 13], and recently shown also by a E3 ubiquitin ligase missing of a typical E3 domain, such as PCAF [14, 15]. The events induced by these E3 ligases lead to proteasome-dependent proteolytic cleavage of Gli2 and Gli3 factors [10, 11], or massive degradation, especially for the Gli1 protein [9, 12, 13].

This chapter analyzes methods for establishing whether a Gli protein (using Gli1 as an example) is ubiquitylated both in vivo and in vitro, as well as protocols to determine whether the covalently linked ubiquitin to lysine residues of target proteins occurs as polyubiquitin chain or monoubiquitylation or multi-monoubiquitylation events.

2 Materials

Prepare all solutions using ultrapure water (deionized water to attain a sensitivity of 18 MΩ cm at 25 °C). All reagents are used following manufacturer's instructions.

2.1 Cell Culture

1. HEK293T cells.

2. HEK293T culture medium: DMEM plus 10 % fetal bovine serum (FBS), L-glutamine and antibiotics.

3. Human D283 medulloblastoma cells.

4. D283 culture medium: DMEM plus 20 % FBS, L-glutamine, and antibiotics.

5. MG132 (Calbiochem, Nottingham, UK), dissolved in DMSO (Dimethyl sulfoxide) and stored at –20 °C. HEK293T cells are treated for 4 h with 50 μM proteasome inhibitor.

2.2 Transfection Reagents

1. Lipofectamine 2000 (Invitrogen, Eugene, OR, USA).

2. Opti-MEM (EuroClone, Milan, Italy).

2.3 Vectors

1. Flag- and HA-tagged ubiquitin.

2. Flag-tagged UbK29/48/63R (provided by Dr I. Dikic).

3. GFP-, Flag-, and HA-tagged Gli1.

4. Myc- and GST-tagged Itch.

5. Flag-tagged PCAF.

2.4   Lysis Buffers

1. Denaturing buffer: 1 % sodium dodecyl sulfate (SDS), 50 mM Tris–HCl pH 7.5, 0.5 mM EDTA, 1 mM dithiothreitol (DTT), and 2 mM *N*-ethylmaleimide (NEM).

2. NETN buffer: 50 mM Tris–HCl pH 7.5, 150 mM NaCl, 5 mM EDTA, 0.1 % NP-40.

3. Triton buffer: 50 mM Tris–HCl pH 7.5, 150 mM NaCl, 0.5 % Triton X-100 (Sigma Aldrich, St. Louis, MO, USA), 50 mM Sodium fluoride (NaF) (Sigma Aldrich), 1 mM EDTA.

4. Ubiquitylation buffer: 25 mM Tris–HCl pH 7.5, 50 mM NaCl, 10 % glycerol, 0.01 % Triton X-100, 1 mM EDTA.

5. Protease and phosphatase inhibitors: 1 mM phenyl-methyl sulfonyl fluoride (PMSF), 2 mM sodium orthovanadate (Na_3VO_4), 10 μg/ml leupeptin, 10 μg/ml pepstatin A, and 10 μg/ml aprotinin.

2.5   SDS Polyacrylamide Gel Components

1. Gels: NuPAGE® Novex® 3–8 % Tris-Acetate Protein Gels, 1.0 mm, 10 well (Life Technologies, Carlsbad, CA, USA).

2. SDS running buffer: Add 50 ml 20× NuPAGE® Tris-Acetate SDS Running Buffer (Life Technologies, Carlsbad, CA, USA) to 950 ml deionized water to prepare 1× SDS running buffer. Fill the upper (200 ml) and lower (600 ml) buffer chambers with 1× running buffer. For reduced samples use 500 μl NuPAGE Antioxidant (Life Technologies, Carlsbad, CA, USA) in the upper buffer chamber.

3. NuPAGE® LDS Sample Buffer (4×) (Life Technologies, Carlsbad, CA, USA). Add to each sample and heat to 100 °C for 5 min.

4. 5 μl SeeBlue® Plus2 Pre-Stained Standard benchmark (Life Technologies).

5. Running conditions: voltage 150 V for 1 h, preferably on ice.

2.6   Immunoprecipitation

1. Denaturing lysis buffer and Triton lysis buffer as described in Subheading 2.4.

2. Mouse monoclonal anti-HA or anti-Flag agarose antibody (Sigma Aldrich, St. Louis, MO, USA), used to immunoprecipitate HA- or Flag-tagged Gli1.

3. Rabbit polyclonal anti-Gli1 antibody (Santa Cruz Biotechnology, Inc., Santa Cruz, CA, USA), used to immunoprecipitate endogenous Gli1.

4. Rabbit polyclonal anti-GFP antibody (Santa Cruz Biotechnology, Inc., Santa Cruz, CA, USA), used to immunoprecipitate GFP–Gli1.

5. 1:1 slurry of protein G agarose or protein A agarose (Santa Cruz Biotechnology, Inc., Santa Cruz, CA, USA), depending on whether a monoclonal or polyclonal antibody is used.

2.7   Immunoblotting Components

1. Pure nitrocellulose membrane, pore size 0.45 μm (PerkinElmer, Waltham, MA, USA).

2. Western blot transfer buffer: Add 50 ml 20× NuPAGE® Transfer Buffer (Life Technologies, Carlsbad, CA, USA) and 10 % methanol (100 ml) to 850 ml deionized water to prepare 1× transfer buffer.

3. Transfer conditions: 400 mA for 2 h 4 °C.

4. Tris-buffered saline (TBS-T; 1×): 150 mM NaCl, 10 mM Tris–HCl pH 8, 0.05 % Tween-20 (Sigma Aldrich). Store at 4 °C.

5. PBS-T: 1× PBS w/o Ca^{2+} and Mg^{2+} plus 0.05 % Tween-20.

6. Blocking solution: 5 % milk in TBS-T or PBS-T depending on antibody used. Store at 4 °C.

7. Antibodies: anti-Flag HRP (Sigma Aldrich, St. Louis, MO, USA), anti-HA HRP, anti-GFP-HRP, and anti-Myc-HRP (Horseradish peroxidase) (Santa Cruz Biotechnology) are prepared in 5 % milk in TBS-T and incubated RT for 1 h. Mouse monoclonal anti-Gli1 (Cell Signaling Technology, Beverly, MA, USA) is prepared in 2.5 % milk in TBS-T and incubated overnight at 4 °C. Mouse monoclonal anti-ubiquitin (Santa Cruz Biotechnology) is prepared in PBS-T and incubated overnight at 4 °C. Mouse monoclonal anti-ubiquitin, VU-1 (LifeSensors, Malvern, PA, USA), is prepared in 5 % milk in TBS-T and incubated overnight at 4 °C. Mouse monoclonal anti-ubiquitin (Life Technologies) is prepared in 5 % milk in TBS-T and incubated overnight at 4 °C. Mouse monoclonal anti-ubiquitin (clone FK1) (Affiniti Research Products Limited, Exeter, UK) is prepared in milk 5 % TBS-T and incubated overnight at 4 °C. HRP-conjugated secondary antibodies are prepared in milk 5 % TBS-T or PBS-T depending on primary antibody used and incubated RT for 30 min.

8. ECL Solution: Western Lightning™ *Plus*-ECL (PerkinElmer, Waltham, MA, USA) consisting of two solutions, Enhanced Luminol Reagent Plus and Oxidizing Reagent Plus used at a 1:1 ratio.

2.8   In Vitro Transcription/ Translation

1. TNT® Coupled Wheat Germ Extract System (Promega, Madison, WI, USA) of which are used: 12.5 μl TNT® Wheat Germ Extract, 1 μl TNT® Reaction Buffer, 0.5 μl TNT® RNA Polymerase (SP6 or T7 depending on vectors used), 0.5 μl amino acid mixture minus methionine. Add to this reaction 0.5 μl RnaseOUT™ Recombinant Ribonuclease Inhibitor (40 u/μl) (Life Technologies), 2 μl L-[^{35}S]methionine

(PerkinElmer, Waltham, MA, USA) corresponding to 0.4 µCi/µl final concentration, 1 µg DNA, and nuclease-free water to arrive to 25 µl final volume.

2. Destaining solution: 5 % methanol and 7 % acetic acid in deionized water.

3. Amplify® (GE Healthcare, Pittsburgh, PA, USA) can be used to increased detection sensitivity.

4. Carestream® Kodak® BioMax® MS film (Sigma Aldrich).

2.9 In Vitro Ubiquitylation Assay

Reaction components:

1. 50 mM Tris–HCl pH 7.5. Store at 4 °C.

2. 5 mM MgCl$_2$. Store at 4 °C.

3. 2 mM ATP (GE Healthcare, Pittsburgh, PA, USA).

4. 0.6 mM DTT. Store at –20 °C.

5. 2.5 µg/µl purified recombinant ubiquitin (Boston Biochem, Cambridge, MA, USA). Store at –80 °C.

6. 5–10 µg purified recombinant Flag-ubiquitin (Sigma Aldrich, St. Louis, MO, USA). Store at –80 °C.

7. 1 µM ubiquitin aldehyde (Boston Biochem, Cambridge, MA, USA). Store at –80 °C.

8. E1 (Boston Biochem, Cambridge, MA, USA). Store at –80 °C.

9. Specific E2 (Boston Biochem, Cambridge, MA, USA) depending on E3 ubiquitin ligase used (e.g., UbcH3 and UbcH5 for βTrCP, UbcH7 for Itch/AIP4, UbcH5b for PCAF). Store at –80 °C.

10. Substrate, in vitro transcribed/translated (IVT): maximum 1–2 µl depending on IVT band intensity.

11. Approximately 300–500 ng recombinant GST-E3 ubiquitin ligase. Store at –80 °C.

12. 2.5 µg/µl methyl ubiquitin (Boston Biochem, Cambridge, MA, USA). Store at –80 °C.

13. Nuclease-free water.

2.10 Recombinant Proteins

1. *Escherichia coli* BL21 strain.

2. Sequence of interest cloned in pGEX vector.

3. LB medium: 1 % Bacto Tryptone, 0.5 % yeast extract, 1 % NaCl, 200 µg/ml ampicillin.

4. Isopropylthiogalactoside (IPTG) (Sigma Aldrich) to a final concentration of 1 mM.

5. GST lysis buffer: 20 mM HEPES pH 7.5, 450 mM NaCl, 2 mM EDTA pH 8, 1 % Triton X-100, 10 % glycerol, 1× protease inhibitor cocktail (Sigma Aldrich).

6. Glutathione Sepharose 4B (GE Healthcare, Pittsburgh, PA, USA) (volume used depending on volume of started preparation).

7. Coomassie solution: 45 % methanol, 10 % acetic acid, 0.25 % Coomassie Brilliant Blue G (Sigma Aldrich).

3   Methods

3.1   In Vivo Ubiquitylation Assay of Ectopic Substrate

1. Co-transfect confluent HEK293T cells seeded in 60 mm dishes with vectors encoding Gli1 (Flag or HA or GFP tagged) and ubiquitin (Flag or HA tagged) in the presence or absence of a specific E3 ubiquitin ligase (i.e., Itch in Fig. 1) (*see* **Notes 1** and **2**). Dilute a maximum of 6 µg total DNA and 2.5 µl Lipofectamine 2000/µg DNA in 200 µl Opti-MEM, incubate 15 min at RT, and add directly into the medium.

2. Twenty-four hours after transfection, treat the cells with the proteasome inhibitor MG132 at final concentration of 50 µM for 4 h (*see* **Note 3**) and then stop, removing the medium and washing the cells 2–3 times with PBS w/o Ca^{2+} and Mg^{2+}.

3. Scrape the cells in PBS w/o Ca^{2+} and Mg^{2+} and resuspend the cell pellets in approximately 100 µl of "denaturing" lysis buffer (*see* **Note 4**) with the addition of 2 mM NEM (*see* **Note 5**) that should be freshly dissolved from a stock concentration of 100 mM.

4. Boil cell lysates at 100 °C for 10 min.

5. Centrifuge lysates at $14,000 \times g$ for 30 min at 4 °C.

6. Dilute the supernatants ten times with NETN buffer.

7. Determine the protein concentration of lysates using the Bradford protein assay.

8. Approximately 1 mg of lysate is immunoprecipitated using 15 µl of mouse monoclonal anti-HA or anti-Flag agarose antibody or 2–4 µg of specific antibody to the protein of interest. Rotate for 2 h or overnight at 4 °C.

9. The following day, wash 30 µl of a 1:1 slurry of protein G or protein A agarose (depending on whether a monoclonal or polyclonal antibody is used) two times with lysis buffer and then add to the lysates. Rotate for 1 h at 4 °C.

10. Wash the agarose pellets five times with NETN buffer and then resuspend the dry beads in 5× loading buffer. Boil at 100 °C for 5 min.

11. Resolve immunoprecipitated proteins by SDS-PAGE and transfer to a nitrocellulose membrane.

12. Block the membrane with 5 % milk in TBS-T for 40 min at RT.

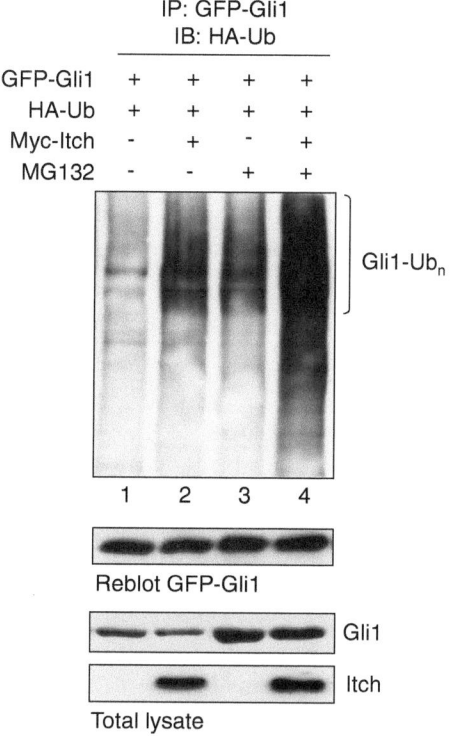

IP: GFP-Gli1
IB: HA-Ub

GFP-Gli1	+	+	+	+
HA-Ub	+	+	+	+
Myc-Itch	-	+	-	+
MG132	-	-	+	+

Gli1-Ub$_n$

1   2   3   4

Reblot GFP-Gli1

Gli1

Itch

Total lysate

Fig. 1 Analysis of ectopic Gli1 ubiquitylation in vivo. HEK293T cells expressing HA-tagged ubiquitin and GFP-tagged Gli1, in the presence (*lanes 2* and *4*) or absence (*lanes 1* and *3*) of Myc-tagged Itch, were treated with proteasome inhibitor MG132 (*lanes 3* and *4*) or DMSO only as control (*lanes 1* and *2*) for 4 h before lysis. Cell extracts were subjected to immunoprecipitation (IP) with anti-rabbit GFP, followed by immunoblot (IB) with anti-mouse HA to detect conjugated HA-Ub (*upper panel*). The blot was reprobed (Reblot) with anti-mouse GFP to detect the efficiency of Gli1 immunoprecipitation. Gli1 and Itch protein levels in total cell lysate are showed in the *bottom panel*. The increase of ubiquitylation and Gli1 protein levels in the presence of MG132 (*lane 3* and *4*) clearly shows that Gli1 is subjected to ubiquitylation processes followed by proteasome-dependent degradation events

13. Incubate with monoclonal anti-Flag® M2-Peroxidase (HRP) or monoclonal anti-HA HRP for Flag- or HA-tagged ubiquitin in 5 % Milk in TBS-T (1:1,000 dilution) for 1 h at RT (*see* **Note 6**). Verify immunoprecipitation efficiency by performing a WB to detect tagged-Gli1 with monoclonal anti-HA or anti-Flag or anti-GFP HRP (1:1,000 dilution) for 1 h at RT.

3.2 In Vivo Ubiquitylation Assay of Endogenous Substrate

1. Use confluent cells (*see* **Notes 7** and **8**).

2. When cells are 80–90 % confluent, treat them with 50 μM MG132 or DMSO only as control for 4 h (*see* Subheading 3.1).

3. Lyse the cells immediately, avoiding freeze them at –80 °C. Resuspend the cell pellets in approximately 100–200 μl of

"denaturing" lysis buffer with the addition of 2 mM NEM (*see* Subheading 3.1).

4. Boil each cell lysate at 100 °C for 10 min.

5. Centrifuged at 14,000×*g* for 30 min at 4 °C.

6. Dilute the supernatant ten times with NETN buffer.

7. Determine the protein concentration of lysates using the Bradford protein assay.

8. Immunoprecipitate if possible more than 1–2 mg of lysate using 2–4 μg of rabbit polyclonal anti-Gli1 H-300 antibody. Rotate overnight at 4 °C.

9. The day after, wash 30 μl of a 1:1 slurry of protein G or protein A agarose (depending on whether a monoclonal or polyclonal antibody is used) two times with lysis buffer and then add to the lysate. Rotate for 1 h at 4 °C.

10. Wash the agarose pellets five times with NETN buffer and then resuspend the dry beads in 5× loading buffer. Boil at 100 °C for 5 min.

11. Resolve immunoprecipitated proteins by SDS-PAGE and transfer on a nitrocellulose membrane.

12. Block the membrane with 5 % milk in PBS-T for 40 min at RT.

13. Incubate the membrane with mouse monoclonal anti-ubiquitin prepared in PBS-T and incubated (1:500 dilution) overnight at 4 °C. To verify immunoprecipitation efficiency WB to detect endogenous Gli1 is performed with mouse monoclonal anti-Gli1 prepared in 2.5 % milk in TBS-T and incubated (1:500) overnight at 4 °C.

3.3 In Vitro Ubiquitylation Assay of a ^{35}S-Labeled Substrate

1. Generate labeled substrate of interest (i.e., HA-Gli1) by in vitro transcription/translation using TNT® Coupled Wheat Germ Extract System (*see* **Note 9**). Prepare a mix of 12.5 μl TNT® Wheat Germ Extract, 1 μl TNT® Reaction Buffer, 0.5 μl TNT® RNA Polymerase (*see* **Note 10**), 0.5 μl Amino Acid Mixture Minus Methionine. Add to this reaction 0.5 μl RnaseOUT™ Recombinant Ribonuclease Inhibitor (40 u/μl), 2 μl L-[^{35}S]methionine (*see* **Note 11**) corresponding to 0.4 μCi/μl final concentration, 1 μg DNA and nuclease-free water to 25 μl final volume. Incubate the reaction at 30 °C for 90 min.

2. Spin the tubes and store the in vitro transcribed/translated (IVT) mix (minus 2 μl) at 20 °C (*see* **Note 12**).

3. Prepare 2 μl of IVT with 5× loading buffer and boil to 100 °C for 5 min.

4. Load and run the sample prepared in **step 3** on SDS-PAGE (*see* **Note 13**).

5. Wash the gel for 30 min at RT with a destaining solution of 5 % methanol and 7 % acetic acid in deionized water.

6. Remove destaining solution and cover the gel with Amplify® for 30 min at RT (*see* **Note 14**).

7. Dry the gel on Whatman 3MM papers for 1 h at 80 °C.

8. Expose to autoradiography overnight using Carestream® Kodak® BioMax® MS film.

9. The day after, verify the in vitro translation band intensity of substrate and decide the volume of IVT to use for ubiquitylation in vitro reaction (*see* **Note 15**).

10. Thaw the following purified recombinant proteins: ubiquitin, ubiquitin aldehyde, E1, E2, GST-E3 ubiquitin ligase, methyl ubiquitin (*see* **Note 16**).

11. Prepare a mix of the following reagents for all experimental points: 50 mM Tris–HCl pH 7.5 (store at 4 °C), 5 mM MgCl$_2$ (store at 4 °C), 2 mM ATP freshly prepared, 0.6 mM DTT (store at –20 °C), 2.5 µg/µl ubiquitin (store at –80 °C), 1 µM ubiquitin aldehyde (store at –80 °C), 1.5 ng/µl E1 (store at –80 °C), 10 ng/µl specific E2 (store at –80 °C) depending on E3 ubiquitin ligase used (i.e., UbcH7 for Itch/AIP4) (*see* **Note 17**), 2.5 µg/µl methyl ubiquitin (*see* **Note 18**) and nuclease-free water.

12. Aliquot the above reaction (8–9 µl) for each experimental point.

13. Add approximately 500 ng recombinant GST-E3 ubiquitin ligase protein (*see* **Note 19**).

14. Add substrate IVT: maximum 1–2 µl depending on IVT band intensity.

15. Add 5× loading buffer immediately in 0 min time points (control) and boil for 1 min at 100 °C, then store at –20 °C until gel loading (*see* **Note 20**).

16. All the other experimental points are heated at 30 °C for the desired times, depending on the substrate studied (*see* **Note 21**).

17. Finish each reaction adding 5× loading buffer, and boiling for 1 min at 100 °C.

18. Load and run the samples prepared in **steps 15** and **17** on SDS-PAGE.

19. Wash the gel for 30 min at RT with a destaining solution of 5 % methanol and 7 % acetic acid in deionized water.

20. Remove destaining solution and cover the gel with Amplify® for 30 min at RT.

21. Dry the gel on Whatman 3MM papers for 1 h at 80 °C.

22. Expose to autoradiography overnight using Carestream® Kodak® BioMax® MS film.

3.4 Ubiquitylation In Vitro Assay of a Transfected Substrate

1. Co-transfect confluent HEK293T cells in 100 mm dishes with empty vector, used as negative control or vector encoding for substrate of interest (i.e., HA-tagged Gli1) or other proteins involved in ubiquitylation reaction, such as components of SCF complex (i.e., Cullin). Prepare a mix of 2.5 μl Lipofectamine 2000/μg DNA (10–15 μg of total DNA) in Opti-MEM, and after 15 min RT, add it directly into the medium.

2. Twenty-four hours after transfection, treat the cells with proteasome inhibitor MG132 at final concentration of 50 μM for 4 h and then remove the medium, and wash the cells two times with PBS w/o Ca^{2+} and Mg^{2+}.

3. Lyse the cells immediately, avoiding freezing them at –80 °C. Add 500 μl of Triton buffer with the addition of protease inhibitors to each 100 mm dish (*see* Subheading 2.4).

4. Keep the lysate on ice for 30 min and the centrifuge at $20,000 \times g$ for 30 min 4 °C.

5. Transfer the supernatant to a new tube and determine protein concentration by Bradford protein assay.

6. Immunoprecipitate, for each experimental point, approximately 1 mg or more of protein lysate adding, for HA substrate, 15 μl of anti-HA agarose beads. Rotate overnight at 4 °C.

7. The day after, wash the agarose beads five times with 500 μl Triton buffer, rotating sample each time for 10 min at 4 °C.

8. Wash two times with 500 μl of ubiquitylation buffer. After the last wash, remove all buffer except the last ~20 μl (1:1 slurry supernatant) and proceed with ubiquitylation reaction.

9. Add to each sample the following ubiquitylation reaction mix: 50 mM Tris–HCl pH 7.5, 5 mM $MgCl_2$, 2 mM ATP, 0.6 mM DTT, 1 μM ubiquitin aldehyde, 40 ng E1, 300 ng specific E2 (i.e., UbcH7 for Itch/AIP4), 5 μg recombinant Flag-ubiquitin, 300–500 ng E3 ubiquitin ligase enzyme (*see* **Note 22**), and nuclease-free water.

10. Immediately add 5× loading buffer in 0 min time points (control) and boil for 1 min at 100 °C, then store at –20 °C until gel loading (*see* **Note 20**).

11. All the other experimental points are heated at 30 °C for the desired times, depending on substrate studied.

12. Stop each reaction adding 5× loading buffer, boil 1 min at 100 °C, and store at –20 °C until gel loading.

13. Load and run the samples prepared in [10, 12] on SDS-PAGE.

14. Transfer on a nitrocellulose membrane.

15. Block the membrane with 5 % milk in TBS-T for 40 min at RT.

16. Perform Western blotting (WB) for ubiquitin using a 1:1,000 dilution of anti-Flag HRP antibody diluted in TBS-T and incubated 1 h at RT.

3.5 Preparation of Recombinant E3 Ligases

1. Transform pGEX-E3 ubiquitin ligase (i.e., pGEX-Itch) in an appropriate *Escherichia Coli* strain (i.e., BL21).

2. Inoculate 50 ml LB medium (*see* Subheading 2.10) with transformants and grow culture overnight.

3. Use the overnight culture to inoculate 400 ml LB medium and grow culture at 37 °C with shaking until an OD_{600} of 0.4–0.6.

4. Induce expression of the fusion protein by adding IPTG to a final concentration of 1 mM to the bacteria culture.

5. Grow culture for additional 3–5 h and then centrifuge at $3,500 \times g$ for 15 min at 4 °C. Store pellet at –80 °C or proceed with lysis.

6. Resuspend the pellet in 8 ml of GST lysis buffer.

7. Vortex and incubate 20 min on ice.

8. Sonicate the samples for three times, transfer the lysate in 1.5 ml tubes, and centrifuge at $20,000 \times g$ for 30 min at 4 °C.

9. Wash 400 μl Glutathione Sepharose 4B one time with 500 μl GST lysis buffer and two times with PBS 1× w/o Ca^{2+} and Mg^{2+}. Resuspend Glutathione Sepharose 4B in a ratio 1:1 with PBS 1× w/o Ca^{2+} and Mg^{2+}.

10. Transfer the supernatants from **step 8** into new 1.5 ml tubes and add 50 μl of prewashed Glutathione Sepharose 4B to each tube.

11. Mix suspension by rotation for 2 h at 4 °C.

12. Wash each pellet five times with 500 μl GST lysis buffer.

13. Pool all pellets into one tube and resuspend pellet in a 1:1 ratio with PBS 1× w/o Ca^{2+} and Mg^{2+} plus protease inhibitors.

14. Analyze expression by SDS-PAGE using 10 μl of your sample plus 2 μl of 5× Laemmli buffer. Detect proteins by Coomassie blue staining.

15. Destain gel with destaining solution (5 % methanol and 7 % acetic acid) and dry 1 h at 80 °C.

4 Notes

1. It is possible to test not only the basal ubiquitylation of substrate (in 1:1 ratio of transfected ubiquitin and substrate vectors) but also in the presence of a specific E3 ubiquitin ligase vector, usually used in 1:1 ratio with substrate vector.

2. It is possible to verify whether a substrate is polyubiquitylated or multi-monoubiquitylated by transfecting the cells with a vector encoding for ubiquitin wild-type protein (Ub wt) or a

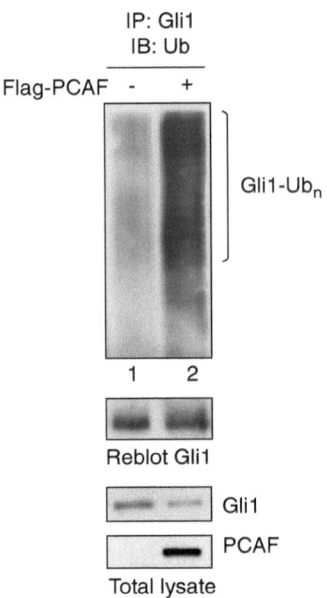

Fig. 2 Analysis of endogenous Gli1 ubiquitylation in vivo. Human D283 medulloblastoma cells were transfected with Flag-tagged PCAF (an acetyltransferase with intrinsic E3-ubiquitin ligase activity) (*lane 2*) or empty vector (*lane 1*). Cell extracts were subjected to IP with anti-rabbit Gli1 antibody followed by IB with anti-mouse Ub antibody to detect conjugated ubiquitin. The blot was reprobed with anti-mouse Gli1 to detect the efficiency of IP. Lower Gli1 protein levels in total cell lysate observed in the presence of PCAF indicate that this ubiquitylation process is followed by proteolytic processing

vector encoding for ubiquitin [mono]substrate mutant UbK29/48/63R (Ub mut) (Fig. 2).

3. Many ubiquitylated proteins are rapidly degraded by proteasomes, making this transient event difficult to detect. For this reason it is useful to treat cells with an inhibitor of the proteasome to preserve the ubiquitylated forms. Moreover, the addition of MG132 or only solvent (DMSO) as control is useful to verify whether the substrate is subjected to proteasome degradation. In this case, in the presence of MG132, a higher accumulation of polyubiquitylated form than control is observed (Fig. 1). Generally, treating cells with the proteasome inhibitor MG132 for 2–4 h is sufficient to block the catalytic activity of the proteasome without resulting in toxicity for the cells. Sometimes it can occur that the inhibition with MG132 is partial, thus some residual degradation products are still detected. This event is indicated by the formation of a ladder at lower size of substrate MW (Fig. 1). It can be useful to add MG132 (at the same concentration indicated in Subheading 2.1) to the lysis buffer.

4. It is necessary to denature the extract before purification of the target protein to demonstrate that the protein of interest is itself ubiquitylated and not merely binding to additional copurified proteins that are ubiquitylated.

5. The presence of deubiquitylating enzymes (isopeptidases) in the cell extract, which can remove the ubiquitin molecules from the protein of interest, could prevent detection of the ubiquitylated species. These problems can be avoided by preparing the extraction buffer with N-ethylmaleimide (NEM), which blocks isopeptidases and the critical cysteine residue present in the active site of most deubiquitylating enzymes.

6. To detect ubiquitylation chains, it is possible to use several antibodies that allow to recognize and to distinguish poly-, multi-mono-, or monoubiquitylated protein, as well as free ubiquitin. Moreover some antibodies recognize K48-, K63-, and K11-linkages as well as linear ubiquitin chains (i.e., anti-ubiquitin antibody, mouse monoclonal, VU-1 LifeSensors).

7. Use cell lines in which a good expression of substrate of interest is known. Use for each experimental point 100 or 150 mm dishes to obtain high concentration of protein lysate.

8. If you want to test if the ubiquitylation of the endogenous substrate is mediated by a specific E3 ubiquitin ligase, you can transfect cells with a vector encoding for E3 ubiquitin ligase of interest (Fig. 3). In this case, it is appropriate using 60–70 % confluent cells and following transfection protocol as indicated in Subheading 2.2.

9. For ubiquitylation in vitro experiment, it is preferable to use Wheat Germ Extract rather than a rabbit reticulocyte lysate because its enzymes could give some background ubiquitylation of the target protein.

10. Use SP6 or T7 depending on promoter of vectors used.

11. Use methionine L-$[^{35}S]$-50 mM tricine, 10 mM β-mercaptoethanol; store at –20 °C.

12. Prepare small aliquots (2 μl) of in vitro transcribed/translated (IVT) and store at –20 °C to avoid repeated freezes and thaws.

13. Choose percentage of polyacrylamide and time of run depending on MW of protein of interest.

14. Amplifying fluorographic reagent can increase sensitivity of detection and significantly reduce exposure times required for weak beta emitters.

15. Use a maximum of 1–2 μl of IVT reaction (if the efficiency of in vitro translation is high, use also a dilution, for example, 0.2–0.5 μl IVT in a final volume of 2 μl).

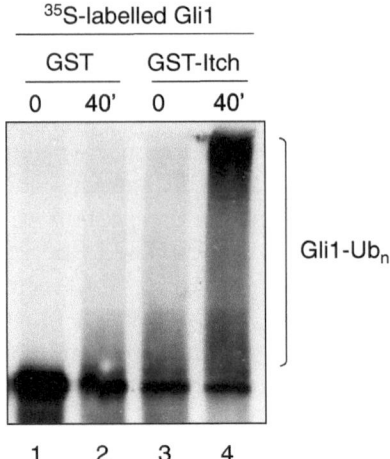

Fig. 3 Analysis of Gli1 ubiquitylation in vitro. In vitro transcribed/translated 35S-labeled HA-tagged Gli1 was subjected to in vitro ubiquitin ligase assay in the presence of purified recombinant ubiquitin, E1, E2 (UbcH7), and in the presence (*lanes 3* and *4*) or absence of recombinant GST-Itch (*lanes 1* and *2*). Samples were incubated at 30 °C for 40 min except that in *lanes 1* and *3*, those were immediately added to sample buffer. ^{35}S-labeled Gli1 ubiquitylation signal was revealed by SDS-PAGE and autoradiography

16. Store all ubiquitin system reagents at −80 °C in small aliquots (maximum 1–2 μl), and don't refreeze after use (they are very susceptible to freeze–thaw cycles).

17. All E3 ubiquitin ligases work in association with a specific E2 ubiquitin-conjugating enzyme, so the choice of E2 to be used in ubiquitylation reaction depends on the E3 ligase system we are studying.

18. The use of methyl ubiquitin is optional because it is used only to verify if the signal is due to polyubiquitylation events: methylated ubiquitin, in fact, is chemically modified to block all of its free amino groups preventing the formation of polyubiquitin chains. The signal in this case appears as lower-molecular-weight ubiquitylated species that show increased stability because they are not efficiently recognized by proteasomes.

19. The recombinant GST-E3 ubiquitin ligase protein used can be homemade (using standard purification protocol of GST-fusion protein from *E. coli* or baculovirus-infected cells) or purchased. When E3 ubiquitin ligases of the RING finger family are used (i.e., F-box protein as β-TrCP), it might need the use of additional enzymes that catalyze posttranslational events on the substrate, such as phosphorylation, which

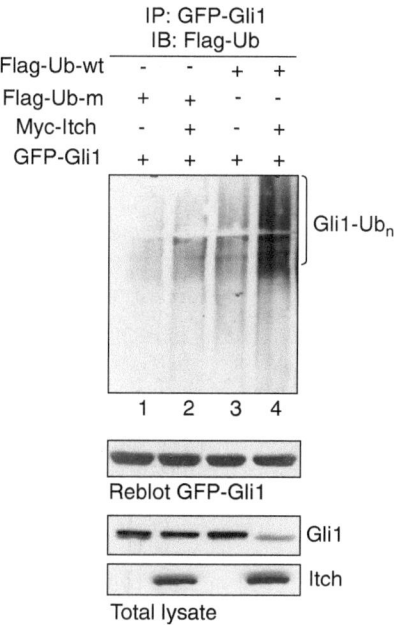

	1	2	3	4
Flag-Ub-wt	-	-	+	+
Flag-Ub-m	+	+	-	-
Myc-Itch	-	+	-	+
GFP-Gli1	+	+	+	+

IP: GFP-Gli1
IB: Flag-Ub

Gli1-Ub$_n$

Reblot GFP-Gli1

Gli1

Itch

Total lysate

Fig. 4 Analysis of Gli1 ubiquitylation in vivo using ubiquitin wild type (Ub wt) or ubiquitin mutant (Ub mut) to test Gli1 polyubiquitylation. HEK293T cells were co-transfected with GFP-tagged Gli1, Flag-tagged ubiquitin (Flag-Ub-wt), or Flag-tagged ubiquitin [mono]substrate (Flag-Ub-mut: Flag-UbK29/48/63R) in the presence or absence of Myc-tagged Itch. Twenty-four hours after transfection, cell lysates were subjected to IP using anti-rabbit GFP antibody followed by IB with anti-mouse Flag to detect conjugated Flag-Ub (*upper panel*). The blot was reprobed (Reblot) with anti-mouse GFP to detect the efficiency of Gli1 immuno-precipitation (*lower panel*). Western blot analysis of Gli1 and Itch levels in total cell lysates is shown in the *bottom panels*. Reduced levels of Gli1 ubiquitylation in the presence of Ub mut, lacking lysines 29, 48, and 63 that are implicated in polyubiquitin chain formation, indicate that Itch induces Gli1 polyubiquitylation

represent the priming signal necessary for this family of E3 ubiquitin ligase to recognize its substrate.

20. This step is necessary to avoid background signal that could occur if the reaction remains at RT for several minutes.

21. To verify the effect of the specific E3 ubiquitin ligase on a substrate, several experimental points in the presence or absence of E3 ubiquitin ligase (to verify the possible background of system) at different time points (i.e., 0, 20, 40 min, or more), depending on substrate studied, are needed (Fig. 4).

22. In the ubiquitylation reaction mix, the E3 ubiquitin ligase used can be a GST-fused recombinant protein (purchased or homemade) (in this case add about 300–500 ng) or an in vitro transcribed/translated (IVT) protein (in this case add about 1–2 μl of the IVT reaction).

Acknowledgment

This work was supported by AIRC (Associazione Italiana Ricerca Cancro) grant #IG14723, the Ministry of University and Research (PRIN project), IIT (Italian Institute of Technology) and Pasteur Institute/Cenci-Bolognetti Foundation.

References

1. Welchman RL, Gordon C, Mayer RJ (2005) Ubiquitin and ubiquitin-like proteins as multifunctional signals. Nat Rev Mol Cell Biol 6:599–609

2. Aberger F, Ruiz I, Altaba A (2014) Context-dependent signal integration by the GLI code: the oncogenic load, pathways, modifiers and implications for cancer therapy. Semin Cell Dev Biol 33:93–104

3. Ruiz i Altaba A, Sanchez P, Dahmane N (2002) Gli and hedgehog in cancer: tumours, embryos and stem cells. Nat Rev Cancer 2:361–372

4. Gulino A, Di Marcotullio L, Canettieri G, De Smaele E, Screpanti I (2012) Hedgehog/Gli control by ubiquitination/acetylation interplay. Vitam Horm 88:211–227

5. Jiang J (2006) Regulation of Hh/Gli signaling by dual ubiquitin pathways. Cell Cycle 5:2457–2463

6. Shabek N, Chiechanover A (2010) Degradation of ubiquitin the fate of the cellular reaper. Cell Cycle 9:523–530

7. Suryadinata R, Roesley SNA, Yang G, Sarcevic B (2014) Mechanism of generating polyubiquitin chains of different topology. Cells 3:674–689

8. Hershko A, Ciechanover A (1998) The ubiquitin system. Annu Rev Biochem 67:425–479

9. Huntzicker EG, Estay IS, Zhen H, Lokteva LA, Jackson PK, Oro AE (2006) Dual degradation signals control Gli protein stability and tumor formation. Genes Dev 20:276–281

10. Pan Y, Bai CB, Joyner A, Wang B (2006) Sonic hedgehog signaling regulates Gli2 transcriptional activity by suppressing its processing and degradation. Mol Cell Biol 26:3365–3377

11. Wang B, Li Y (2006) Evidence for the direct involvement of βTrCP in Gli3 protein processing. Proc Natl Acad Sci U S A 103:33–38

12. Di Marcotullio L, Ferretti E, Greco A, De Smaele E, Po A, Sico MA et al (2006) Numb is a suppressor of Hedgehog signalling and targets Gli1 for Itch-dependent ubiquitination. Nat Cell Biol 8:1415–1423

13. Di Marcotullio L, Greco A, Mazzà D, Canettieri G, Pietrosanti L, Infante P et al (2011) Numb activates the E3 ligase Itch to control Gli1 function through a novel degradation signal. Oncogene 30:65–76

14. Mazzà D, Infante P, Colicchia V, Greco A, Alfonsi R, Siler M et al (2013) PCAF ubiquitin ligase activity inhibits Hedgehog/Gli1 signaling in p53-dependent response to genotoxic stress. Cell Death Differ 20:1688–1697

15. Infante P, Canettieri G, Gulino A, Di Marcotullio L (2014) Yin-Yang strands of PCAF/Hedgehog axis in cancer control. Trends Mol Med 20:416–418

Chapter 13

Determination of Acetylation of the Gli Transcription Factors

Sonia Coni, Laura Di Magno, and Gianluca Canettieri

Abstract

The Gli transcription factors (Gli1, Gli2, and Gli3) are the final effectors of the Hedgehog (Hh) signaling and play a key role in development and cancer. The activity of the Gli proteins is finely regulated by covalent modifications, such as phosphorylation, ubiquitination, and acetylation. Both Gli1 and Gli2 are acetylated at a conserved lysine, and this modification causes the inhibition of their transcriptional activity. Thus, the acetylation status of these proteins represents a useful marker to monitor Hh activation in pathophysiological conditions. Herein we describe the techniques utilized to detect in vitro and intracellular acetylation of the Gli transcription factors.

Key words Gli, Acetylation, Hedgehog

1 Introduction

Hedgehog signaling is a master regulator of tissue development in mammals, and deregulation of the signaling often leads to severe developmental abnormalities and/or tumorigenesis. Gli (glioma-associated oncogene) proteins are the final effectors of Hedgehog pathway, and posttranslational modifications such as phosphorylation, ubiquitination, and SUMOylation are crucial regulated checkpoints of their activity [1].

We have identified the acetylation of Gli1 and Gli2 (two isoforms of the Gli transcription factors) and demonstrated that this covalent modification is a key inhibitory event in the Hedgehog signaling pathway [2, 3]. Acetylation/deacetylation processes are regulated by histone acetyltransferases (HATs) and histone deacetylases (HDACs), which covalently modify histone or nonhistone substrates [4].

Acetylation is mediated by HATs, which catalyze the transfer of acetyl groups to the ε-amino group of lysine residues. HAT enzymes are organized in three major groups: the GCN5-related *N*-acetyltransferases (GNATs), the E1A-associated protein of

Natalia A. Riobo (ed.), *Hedgehog Signaling Protocols*, Methods in Molecular Biology, vol. 1322,
DOI 10.1007/978-1-4939-2772-2_13, © Springer Science+Business Media New York 2015

300 kDa (p300)/CREB-binding protein (CBP), and MYST proteins [5].

Acetylation is reverted by HDACs, a group of enzymes containing a catalytic domain responsible for the removal of the acetyl groups. HDACs are organized into four classes, depending on sequence identity and domain organization. Class I (HDAC1, 2, 3, and 8), class II (HDAC4, 5, 6, 7, 9, and 10), and class IV (HDAC11) are zinc-dependent HDACs, while class III (SIRT1–7) deacetylases are NAD+ dependent [6]. In this brief protocol, we will describe two assays for the analysis of Gli acetylation. The first one allows detection of Gli acetylation in mammalian cells by using standard immunoprecipitation and western blotting (intracellular acetylation assay); the second one (in vitro acetylation assay) is performed with purified Gli proteins (from mammalian cells) and the HAT domain (from *E. coli*), and acetylation is revealed by western blotting or autoradiography.

2   Materials

2.1 Expression of Gli Proteins, P300, and Immunoprecipitation (IP)

1. Human embryonic kidney (HEK) 293T cultured cells (ATCC).

2. DMEM (Dulbecco's Modified Eagle Medium) supplemented with 10 % fetal bovine serum, antibiotics (penicillin–streptomycin), and 1 mM L-glutamine.

3. pcDNA3 Flag-Gli1 or pcDNA3 Myc-Gli2 or pcDNA3 HA-p300 eukaryotic expression plasmids.

4. Lysis buffer Radioimmunoprecipitation (RIPA): 0.5 % sodium deoxycholate, 50 mM Tris–HCl pH 7.6, 1 % NP40, 0.1 % Sodium Dodecyl Sulphate (SDS), 140 mM NaCl, 5 mM Ethylenediaminetetraacetic acid (EDTA) pH 8, 5 mM sodium pyrophosphate, 5 mM sodium butyrate, supplemented with protease inhibitors (1 mM Phenylmethanesulfonylfluoride (PMSF), 2 μg/ml leupeptin, 1 μg/ml aprotinin, 1 μg/ml pepstatin).

5. Anti-Flag M2 affinity gel (Sigma-Aldrich).

6. Anti-Myc affinity gel (Sigma-Aldrich).

7. Anti-acetyl lysine, rabbit polyclonal antibody (Upstate-Cell Singnaling).

8. Mouse Flag (Sigma-Aldrich) and mouse anti-Myc (Santa Cruz Biotechnology) antibodies.

2.2 Production of Recombinant GST-HAT

1. *E. coli* BL21 strain (New England Biolabs).

2. 1 M Isopropil-β-D-1-tiogalattopiranoside (IPTG) (isopropyl-β-D-1-thiogalactopyranoside) (Sigma-Aldrich).

3. NETN buffer (20 mM Tris–HCl pH 8, 100 mM NaCl, 1 mM EDTA, 0.5 % NP-40).

4. TST buffer, also known as TBS (50 mM Tris–HCl pH 8, 150 mM NaCl).

5. Elution buffer (50 mM Tris–HCl pH 8, 150 mM NaCl, 25 mM GSH).

6. Reduced L-glutathione (GSH) (Sigma-Aldrich).

7. Glutathione S-transferase (GST) Sepharose beads (250-μl packed volume (pv)/500-ml bacterial preparation) (Amersham Pharmacia Biotech).

2.3 In Vitro Acetylation Reactions (Cold and Hot) and Coomassie Staining

1. 5× HAT buffer (250 mM Tris–HCl pH 8; 50 % glycerol; 0.5 mM EDTA; 5 mM Dithiothreitol (DTT); 50 mM sodium butyrate).

2. 5 mM acetyl coenzyme A (Sigma-Aldrich).

3. Phosphate buffer saline solution (1× PBS).

4. Coomassie staining solution: 50 % methanol/10 % acetic acid/40 % H_2O with 0.0125 g/l Coomassie R250.

5. Destaining solution: 50 % methanol/10 % acetic acid/40 % H_2O.

6. ^{14}C AcCoA 0.02 mCi/ml (PerkinElmer).

7. Fixing solution: 50 % H_2O/40 % methanol/10 % acetic acid.

8. Enhancer solution (Bio-Rad).

2.4 Western Blotting

1. 8 % SDS polyacrylamide gel.

2. Nitrocellulose 0.45 μm (PerkinElmer).

3. Running buffer solution: 25 mM Tris base, 192 mM glycine, 0.1 % SDS.

4. Blotting buffer solution: 25 mM Tris base, 192 mM glycine, 10 % methanol.

5. Washing buffer Tris-Base-SDS plus Tween solution (TBS-T): 10 mM Tris–HCl (pH 7.5), 100 mM NaCl, 0.1 % Tween-20.

6. Blocking solution: 5 % not-fat dry milk in TBST.

7. Laemmli buffer 2×: 4 % SDS, 10 % 2-mercaptoethanol, 20 % glycerol, 0.004 % bromophenol blue, 0.125 M Tris–HCl, pH 6.8.

8. Benchmark pre-stained protein ladder (Invitrogen).

---

3 Methods

3.1 Cell Culture and Transfection

1. Seed 3.5×10^6 HEK 293T cells 24 h before transfection onto 21-mm^2 dishes for the intracellular acetylation assay, or 7×10^6 cells on 56-mm^2 dishes for the in vitro acetylation assay in DMEM, supplemented with 10 % fetal bovine serum, 1 mM penicillin–streptomycin, and 1 mM L-glutamine at 37 °C, 5 % CO_2.

2. Transfect cells using Lipofectamine Reagent (Invitrogen) at 1:3 ratio with 4 μg total DNA/21-mm^2 dish or 8 μg total DNA/56-mm2 dish. For the *intracellular* acetylation assay, transfect 21-mm^2 dishes with:

   (a) pcDNA3 Flag-Gli1 (or pcDNA3 Myc-Gli2) (2 μg) + pcDNA3 empty vector (2 μg).

   (b) pcDNA3 Flag-Gli1 (or pcDNA3 Myc-Gli2) (2 μg) + pcDNA3-HAp300 (2 μg).

   For the in vitro acetylation assay, transfect cells in 56-mm^2 dishes with pcDNA3 Flag-Gli1 (or pcDNA3 Myc-Gli2) (8 μg DNA/dish).

   Briefly, mix Lipofectamine Reagent with Opti-MEM and incubate for 5 min. Add 500 μl of the mix to each DNA sample and incubate 20 min. Add 500 μl of the total mix to the dishes and adjust the final volume to 2 ml with Opti-MEM for the 21-mm^2 dish or to 5 ml for the 56-mm^2 dish.

3. Incubate the cells for 4 h at 37 °C, 5 % CO_2. Remove the medium and replace with fresh DMEM plus 10 % Fetal Bovine Serum (FBS) and 1 mM L-Glutamine without antibiotics.

4. After 24 h, proceed with lysis for immunoprecipitation.

3.2 Immuno-precipitation

1. Lyse the cells in RIPA buffer (500 μl for each dish 21 or 56 mm^2) supplemented with protease inhibitors as indicated in Subheading 2.1. Incubate lysates on ice for 30 min and then centrifuge 30 min at +4 °C.

2. Collect the supernatant and quantify protein content using the Bradford protein assay (Bio-Rad). Dilute the Bradford solution 1:5 with distilled water. Then add 1 μl of each sample to 1 ml of the Bradford solution and determine sample concentration by spectrometry at $\lambda = 595$ and compare to the blank solution, which contains 1 μl of lysis buffer in 1 ml of Bradford.

3. To purify transfected Gli proteins, transfer 1 mg of total protein lysate into a 1.5 ml tube and add 20 μl of anti-Flag M2/anti-Myc affinity gel (*see* **Note 1**). Incubate samples with gentle rocking for 2 h at +4 °C.

4. Centrifuge the IP samples at $2,500 \times g$ for 5 min at +4 °C. Discard the supernatant and add 800 μl of lysis buffer to the beads (*see* **Note 2**). Mix by flickering the samples and inverting the tubes 4–5 times. Centrifuge again at $2,500 \times g$ for 5 min at +4 °C. Repeat this step five times.

5. After the last wash, elute the samples for the *intracellular acetylation assay* or the in vitro *reaction* (*see* **Note 3**). Add 20 μl of Laemmli buffer to each sample and mix well. Heat the tubes at 100 °C for 5 min and centrifuge at $16,000 \times g$ for 5 min. Collect the supernatant (20 μl per sample).

3.3   Intracellular Acetylation Assay

1. Prepare an 8 % SDS-PAGE gel as follows: for 10 ml of separating gel, mix 4.6 ml H_2O; 2.6 ml of a 30 % acrylamide solution (29:1); 2.6 ml of 1.5 M Tris–HCl pH 8.8; 100 µl 10 % SDS; 100 µl 10 % APS; and 6 µl Tetramethylethylenediamine (TEMED). Pour the gel and let it to polymerize. For 4 ml of stacking gel, mix 2.8 ml H_2O; 0.66 ml of 30 % acrylamide solution (29:1); 0.5 ml of 1.5 M Tris–HCl pH 6.8; 40 µl 10 % SDS; 40 µl 10 % APS; and 4 µl TEMED.

2. Load the eluted IP samples. Perform electrophoresis at 120 V for 1 h in running buffer solution. Transfer gel onto a nitrocellulose membrane (2 h at 400 mA) with the wet-tank method in blotting buffer at +4 °C.

3. After transfer, wash the nitrocellulose membrane with TBS-T for 5 min at room temperature.

4. Incubate in blocking buffer for 1 h at room temperature, followed by incubation with anti-acetyl lysine (upstate Cell Rignaling) primary antibody diluted 1:1,000 in blocking solution overnight.

5. Wash the membrane three times for 5 min each with TBS-T.

6. Incubate with anti-rabbit secondary antibody (Santa Cruz Biotechnology) diluted 1:5,000 in blocking solution for 45 min.

7. Wash the membrane three times for 5 min each with TBS-T.

8. Develop the blot using Luminol-based enhanced chemiluminescence (ECL) (PerkinElmer).

3.4   Production of Bacterially Expressed Recombinant HAT for In Vitro Acetylation Assays

In vitro acetylation assay requires the purification of recombinant GST-HAT expressed in *E. coli* and the purification of Gli proteins overexpressed in mammalian cells (see above) with immunoaffinity chromatography. Using our reagents and technique, we usually obtain 0.2–0.25 µg of purified protein per ml of bacterial culture. Below is the protocol to purify about 100 µg of GST-HAT proteins.

1. Transform *E. coli* BL21 strain with the pGEX-HAT vector (*see* **Note 4**). After addition of the plasmid to the competent cells, incubate on ice for 30 min, place tube in a 42 °C bath for exactly 45 s (heat shock), and let it cool 2 min on ice. Then add 900 µl of LB medium and then incubate for 45 min with shaking at 200–250 rpm at 37 °C. Collect the bacteria by centrifugation at $900 \times g$ for 3 min, plate onto ampicillin LB agar plates and incubate overnight (16 h) at 37 °C.

2. On day 2, inoculate a single colony into 10 ml of LB + ampicillin (50 µg/ml), and incubate with agitation at 37 °C overnight.

3. On day 3, add the 10 ml of the above bacterial culture to 500 ml of LB + ampicillin and shake at 200–250 rpm at 37 °C. Monitor bacterial growth by measuring the OD_{600} and, when

it reaches 0.4 OD (in about 2 h), add 500 μl of 1 M IPTG to the bacterial culture to obtain a final concentration of 1 mM. Incubate for 2 h at 37 °C with agitation. Collect 1 ml of bacterial culture before and After the IPTG induction steps to be loaded on a Coomassie gel (*see* **Note 5**).

4. At the end of the induction, pellet bacteria by centrifugation at 2,700×*g* for 15 min at +4 °C and resuspend in 10 ml of 1× PBS. Centrifuge again at 2,700×*g* and carefully resuspend in 10 ml of NETN lysis buffer, supplemented with protease inhibitors (1 mM PMSF, 2 μg/ml leupeptin, 1 μg/ml apro-tinin, 1 μg/ml pepstatin).

5. Sonicate the preparation three times for 30 s each, incubate 15 min on ice and centrifuge again at 8,000×*g* for 30 min. Keep the supernatant and take a 20 μl sample for a Coomassie gel.

6. To equilibrate the Sepharose beads, wash 250 μl of GST Sepharose beads with 1 ml NETN buffer and centrifuge at 2,400×*g* for 5 min. Resuspend the pelleted beads in 250 μl of NETN buffer (1:1).

7. Add the equilibrated 250 μl GST Sepharose beads to the bac-terial lysate. Incubate the sample in a rotator for 2 h at +4 °C.

8. Centrifuge the samples at 5,000×*g* at +4 °C, discard the super-natant, and wash the beads five times with 10 ml of NETN buffer (40 volumes of the packed beads), by inverting the tube 4–5 times and centrifuging at 5,000×*g* at +4 °C. Perform two additional washes with TST buffer.

9. Add 25 mM GSH to the TST buffer to make the elution buf-fer. Add 150 μl of elution buffer to the pelleted beads and incubate for 10 min at room temperature. Centrifuge at 16,000×*g* for 5 min and collect the supernatant. Repeat the elution step four times and pool the supernatants. Determine the final protein concentration using the Bradford protein assay (Bio-Rad).

10. To check the expression of the GST proteins, load 20 μl of the eluted protein and known quantities of BSA (as control) onto an 8 % SDS-PAGE gel (Fig. 1). Perform electrophoresis at 120 V for 1 h in running buffer solution.

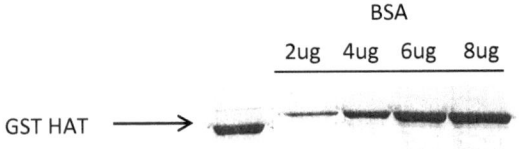

Fig. 1 Coomassie staining of the GST-HAT recombinant protein. 20 μl of eluted purificated protein (first lane on the *left*) is compared to known quantity of BSA for optical quantification. In this experiment the concentration is approximately 0.2 μg/ml

11. Stain the gels with Coomassie solution for 30 min, and wash three times with destaining solution three times (15 min/each) or with distilled water overnight (*see* **Note 6**). Adjust concentration of the purified GST-HAT to 0.2 µg/ml.

3.5 In Vitro Acetylation Reactions

Acetylation of Gli proteins can be detected in vitro by western blotting with the anti-acetyl lysine antibody (*cold* in vitro *reaction*) or by autoradiography of incorporated ^{14}C AcCoA (*hot* in vitro *reaction*).

3.5.1 Cold In Vitro Reaction

1. Prepare two different mixes (HAT mix and negative control without HAT) in a final volume of 200 µl.
   (a) 5× HAT buffer (40 µl) + AcCoA (4 µl) + H_2O (136 µl) (for each sample).
   (b) 5× HAT buffer (40 µl) + AcCoA (4 µl) + H_2O (131 µl) + GST-HAT at 0.2 µg/µl (5 µl) (for each sample).

2. Add the mixes directly on the washed beads (20 µl of packed beads) containing immunoprecipitated Gli1 or Gli2 (prepared in Subheading 3.2). Incubate the samples at 30 °C for 1 h (Table 1), mixing the samples every 5 min.

3. Centrifuge the tubes at 2,400×g for 30 min. Discard the supernatant and add 20 µl of Laemmli buffer to each sample. Heat for 5 min at 100 °C and load on 8 % SDS-PAGE gels, and western blot is performed as described above.

3.5.2 Hot In Vitro Reaction

1. Prepare two different mixes (HAT mix and negative control without HAT) in a final volume of 200 µl.
   (a) 5× HAT buffer (40 µl) + [^{14}C]AcCoA (4 µl) + H_2O (136 µl) (for each sample).
   (b) 5× HAT buffer (40 µl) + [^{14}C]AcCoA (4 µl) + H_2O (131 µl) + GST-HAT at 0.2 µg/µl (5 µl) (for each sample).

2. Add the mixes directly on the washed beads (20 µl of packed beads) containing immunoprecipitated Gli1 or Gli2 (prepared in Subheading 3.2). Incubate the samples at 30 °C for 1 h (Table 1), mixing the samples every 5 min.

3. Centrifuge the tubes at 2,400×g for 30 min. Discard the supernatant (*see* **Note 7**) and add 20 µl of Laemmli buffer to each sample. Heat for 5 min at 100 °C and load on 8 % SDS-PAGE. Electrophoresis is performed for 1 h at 120 V (*see* **Note 8**).

4. At the end of the run, wash the gel with the fixing solution for 20 min, then with the enhancer solution for 20 min. Dry the gel at 80 °C for 1 h and expose to a Kodak photographic film at −80 °C. Representative results are shown in Fig. 2.

Table 1
Scheme for the in vitro acetylation reaction

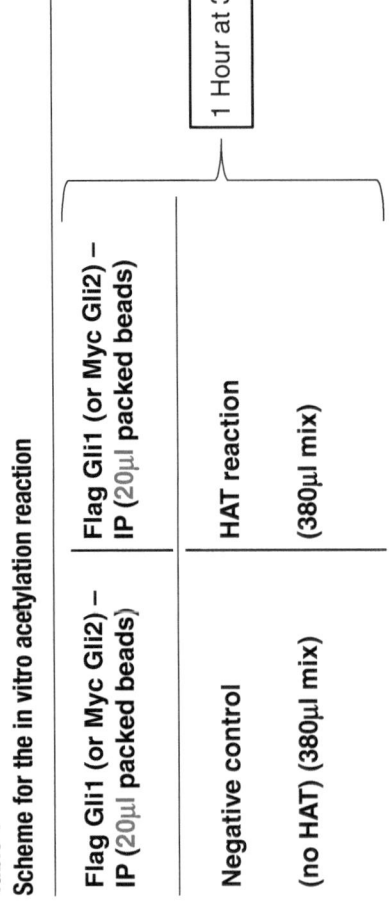

Flag Gli1 (or Myc Gli2) – IP (20µl packed beads)	Flag Gli1 (or Myc Gli2) – IP (20µl packed beads)	
Negative control	HAT reaction	1 Hour at 30°C
(no HAT) (380µl mix)	(380µl mix)	

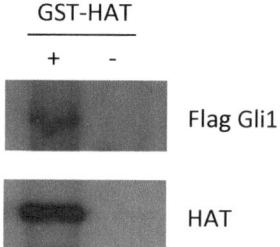

GST-HAT

\+   -

Flag Gli1

HAT

Fig. 2 Hot acetylation assay of Gli1. Autoradiography of acetylated Gli1, detected after 5 days of exposure (*upper panel*). GST-HAT autoacetylation signal is also detected (*bottom panel*)

4 Notes

1. Cut the top of the tips to facilitate pipetting of agarose gel (or Sepharose gel where it is indicated).

2. Forty volumes of washing buffer should be used for each washing.

3. To perform the in vitro *reaction*, beads should be equilibrated two times with the 1× HAT buffer (diluted with H_2O from the 5× HAT stock buffer). In this case, prepare two samples for each point to be analyzed (one for the HAT reaction another one for the negative control—no HAT).

4. *E. coli* BL21 bacteria are available from New England Biolabs and are suitable for the production of recombinant proteins because of their high transformation efficiency $(1–5 \times 10^7$ cfu/μg DNA) and the absence of endogenous proteases.

5. It is advisable to monitor each step of purification of your recombinant protein by setting side an aliquot of bacterial culture or lysate for each step and run it on a Coomassie gel, as described in the text. You can load a gel with the pre-IPTG, post-IPTG, pre-IP beads, and post-IP beads samples that you can collect to check the efficiency of GST purification.

6. The Coomassie gel should be dried in a gel dryer for 1 h on a 3-mm wet paper, covered with a transparent film.

7. Use proper bags as a waste container when you work under the chemical hood to perform the hot acetylation assay.

8. You can use recombinant histones (purchased from New England Biolabs) as positive control for your in vitro acetylation assay reaction.

Acknowledgements

This work was supported by AIRC (Associazione Italiana Ricerca Cancro) grant # IG10610, MIUR FIRB and PRIN projects, Ministry of Health, EU HEALING grant # 238186, Pasteur Institute-Cenci Bolognetti Foundation, and Italian Institute of Technology (IIT), NIH P41 RR011823. Laura Di Magno (LDM) and Sonia Coni (SC) were supported by fellowships from the Pasteur Institute-Cenci Bolognetti Foundation.

All authors disclose no financial conflict of interest that might be construed to influence the results or interpretation of their manuscript.

References

1. Hui C-C, Angers S (2011) Gli proteins in development and disease. Annu Rev Cell Dev Biol 27:513–537

2. Canettieri G, Di Marcotullio L, Greco A, Coni S, Antonucci L, Infante P et al (2010) Histone deacetylase and Cullin3-REN(KCTD11) ubiquitin ligase interplay regulates Hedgehog signalling through Gli acetylation. Nat Cell Biol 12:132–142

3. Coni S, Antonucci L, D'Amico D, Di Magno L, Infante P, De Smaele E et al (2013) Gli2 acetylation at lysine 757 regulates hedgehog-dependent transcriptional output by preventing its promoter occupancy. PLoS One 8:e65718

4. Di Marcotullio L, Canettieri G, Infante P, Greco A, Gulino A (2011) Protected from the inside: endogenous histone deacetylase inhibitors and the road to cancer. Biochim Biophys Acta 1815:241–252

5. Lee KK, Workman JL (2007) Histone acetyltransferase complexes: one size doesn't fit all. Nat Rev Mol Cell Biol 8:284–295

6. Akimova T, Beier UH, Liu Y, Wang L, Hancock WW (2012) Histone/protein deacetylases and T-cell immune responses. Blood 119:2443–2451

Efficient Detection of Indian Hedgehog During Endochondral Ossification by Whole-Mount Immunofluorescence

João Francisco Botelho, Daniel Smith-Paredes, and Verónica Palma A.

Abstract

Endochondral ossification is a process essential for the formation of the vertebrate skeleton. Indian Hedgehog (IHH) is a key regulator of this process. So far, monitoring IHH expression in whole-mount developing skeletal structures has been hampered by the permeability and the opacity of the tissue. Whole-mount preparations require advanced techniques of fixation, clearing, and staining. We describe a reliable method for fixing, immunostaining, and clearing whole-mount developing cartilages that allows for the detection of IHH in the developing skeleton of avian embryos. The fixation process ensures a proper preservation of cellular structures and, especially, the antigenicity of the tissue, allowing the antibody labelling of IHH. This protocol reveals specific cell staining in localized regions of the developing cartilage, facilitating the study of IHH function during key periods of skeletogenesis.

Key words IHH, Collagen, Whole-mount immunofluorescence, Endochondral ossification, Avian embryo

1 Introduction

Most bones develop by an indirect process called endochondral ossification [1–3]. This process begins by the formation of small cartilaginous molds that are replaced by mineralized tissue during development. The early cartilaginous precursors are entirely composed by slow-proliferating stem chondrocytes that start differentiating in the center of the cartilage, becoming flattened and fast-proliferating. Flattened chondrocytes then exit the cell cycle and start growing, turning hypertrophic (Fig. 1), and eventually dying by apoptosis, leaving only the extracellular matrix in the center of the element. This matrix calcifies and is remodelled by action of osteocytes and osteoclasts, respectively [4]. Cartilage growth is achieved by the maintenance of a domain of stem chondrocytes transiting from the stem cell condition to hypertrophy in each epiphysis, a zone called grow plate (Fig. 1a).

Natalia A. Riobo (ed.), *Hedgehog Signaling Protocols*, Methods in Molecular Biology, vol. 1322, DOI 10.1007/978-1-4939-2772-2_14, © Springer Science+Business Media New York 2015

a

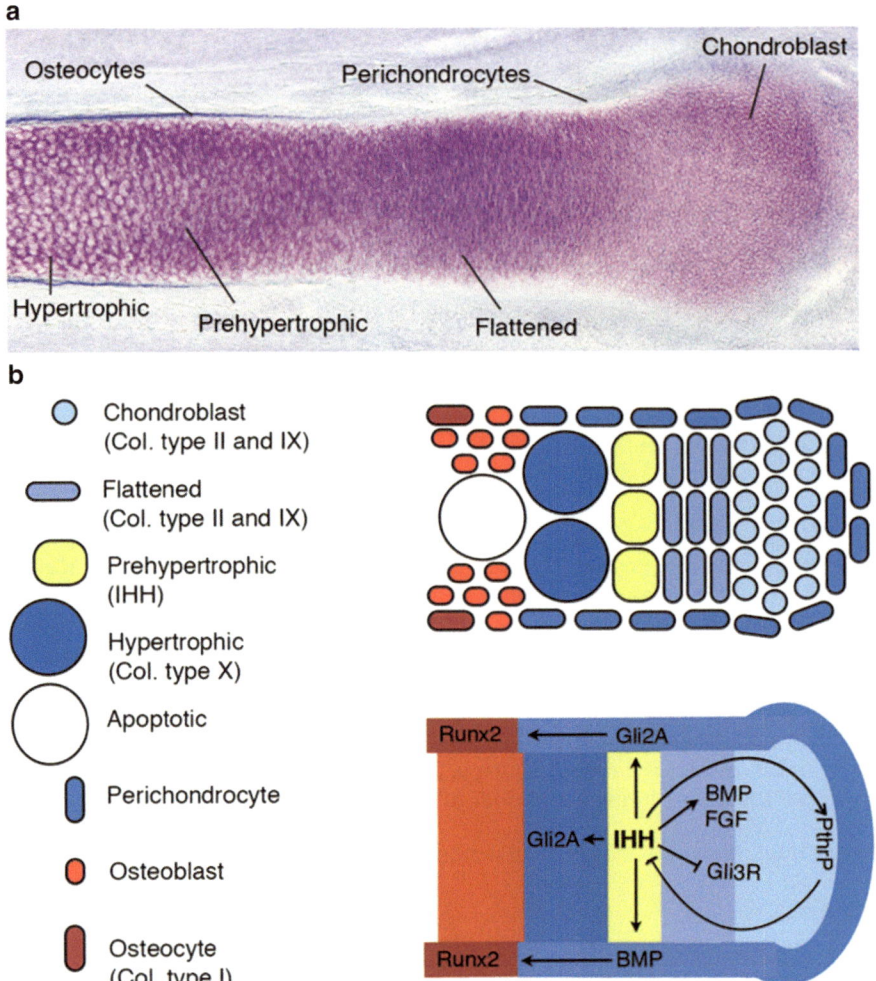

Fig. 1 IHH and Endochondral ossification: (**a**) Histological section of a typical long bone cartilage, depicting the different cell types involved in its development: A population of chondroblasts resides at the ends of the element; as the chondrocytes differentiate, towards the center of the element, they become flatten and thereafter begin to grow; hypertrophic chondrocytes later dye by apoptosis, leaving only the extracellular matrix for calcification. Perichondrial cells around the cartilage generate osteoblasts over the center of the element. (**b**) Each of the different cell types involved in endochondral ossification can be identified, either by their morphology or by the molecular markers they express. A number of proteins have been identified which play roles in regulating bone development. IHH and PTHrP form a feedback loop that regulates the rate of chondrocyte differentiation. The IHH/PTHrP signaling loop also interacts with members of the bone morphogenic protein (BMP) and fibroblast growth factor (FGF) family

Indian Hedgehog (IHH) has been described as being involved in the main molecular pathway controlling different aspects of endochondral ossification [5–10]. During endochondral ossification, IHH molecules secreted by pre-hypertrophic chondrocytes control the pace of proliferation of quiescent chondrocytes [11, 12], the transition to hypertrophic chondrocytes [10, 13], and the production of osteoblasts in the perichondrium [4, 7, 14, 15].

Therefore, IHH is in the crossroad of every aspect of endochondral ossification and understanding its functioning is central to comprehend those processes (Fig. 1b).

In order to determine chondrogenesis at the gross morphological level, Alcian Blue staining is very informative and has been widely used for formalin fixed cartilage tissue (Fig. 2a). Nevertheless, the study of the role of IHH on early endochondral ossification faces some technical hindrances. Due to its water-rich cartilaginous extracellular matrix, skeletal tissue has physical properties that difficult tissue conservation and sectioning. It is also hard to orient the sample in a way to cut an entire element showing its two epiphyses or more than one epiphysis in adjacent elements. Furthermore, cartilages also exhibit sparse cell distribution that results in low signal from assays targeting cellular components. Consequently, much of our current knowledge about IHH function on skeletal development is based on digitally enhanced signal obtained from in situ hybridization on frozen sections showing small parts of each bone.

Most of those hindrances could be surpassed by whole-mount technics as they avoid sectioning and allow easier comparison of parts. Unfortunately, whole-mount visualization is also troublesome because skeletal tissue exhibits very low penetrability, especially after the use of crosslink fixatives (e.g., formalin, paraformaldehyde, glutaraldehyde). In situ hybridization protocols circumvent the problem of penetration on crosslink fixed embryos treating them with Proteinase K, a broad-spectrum protease. But even enzymatic treatments are unable to allow penetration of the large RNA probes in mature cartilages. For this reason, typical in situ hybridization techniques are often employed at earlier stages, as they are not useful at latter stages (e.g., [16]) (Fig. 2b).

Other main inconvenient issue for whole-mount technics is the need to make samples transparent, to allow observation of any tissue located as deep as the skeleton is located among other tissues. The most often employed clearing agents are solvents like methyl salicylate [17] and BABB (benzyl alcohol and benzyl benzoate) [18]. They have several setbacks as they produce tissue deformations, tissue rigidity, high light scattering, are toxic and could damage optical devices. Recently solvent-free solutions of urea [19], fructose [20], and formamide [21] have been proposed as clearing agents for fluorescent samples. They are water-based, allowing safer and easier sample manipulation, and reduce light scattering when compared to solvent-based agents.

We have elaborated a protocol that allows the immunofluorescent staining of IHH protein in whole-mount developing bird cartilages (Fig. 2c). The avian embryonic limb is a classical model for the study skeletal development [22, 23] and its usefulness should be improved by the recent advances on the production of chick transgenic lines [24]. This new protocol avoids crosslink fixation of the tissue employing fast dehydration and cold protein precipitation

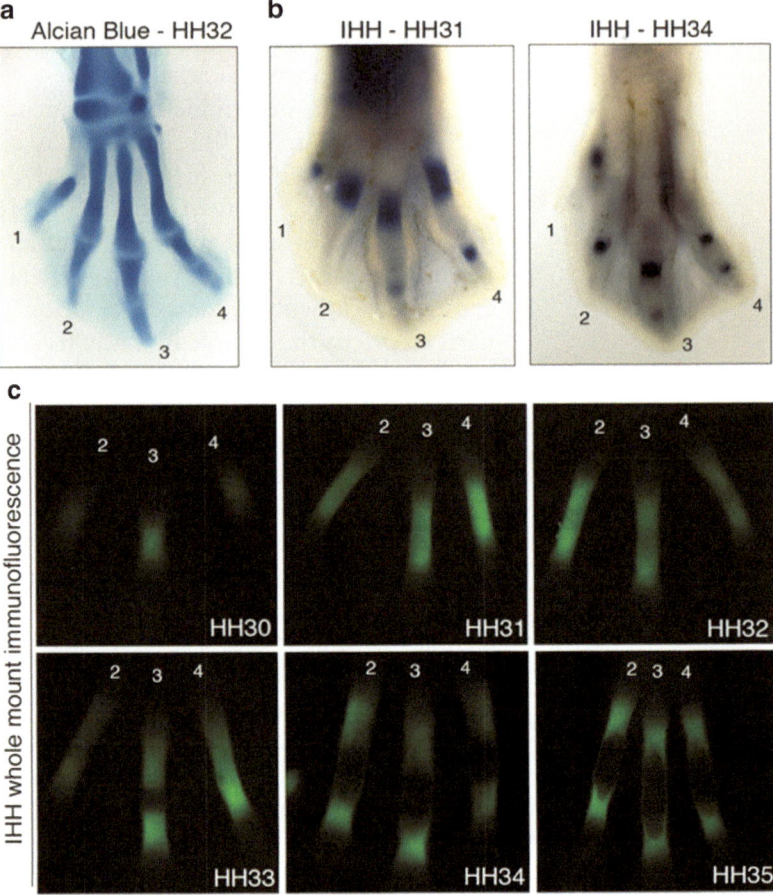

Fig. 2 IHH immunofluorescence: The development of avian metatarsi and phalanges yields a good example of the utility of IHH whole-mount immunofluorescence assay. (**a**) Alcian blue staining reveals the characteristic morphology of the avian hind limb skeleton at early stages (**b**) In situ hybridization is unable to reveal the presence of *IHH* mRNA in the metatarsi at late stages. (**c**) IHH whole-mount immunofluorescence shows how IHH splits in two domains corresponding to the sites where chondrocytes begin to hypertrophy, as development proceeds. In contrast with in situ hybridization or other protocols, penetrability of antibodies is highly enhanced, allowing the visualization of the expression pattern of IHH protein with great detail and at later stages of development

as a fixative method. It increases epitope accessibility through high concentration of detergents and chemical digestion of cartilage by specific enzymatic digestion, while clearing is obtained by a water-based solution of urea. In order to contextualize IHH production, the protocol here is combined with immunofluorescence to collagen subtypes produced by quiescent chondrocytes (collagen types II and IX), hypertrophic chondrocytes (collagen type X), or osteoblasts (collagen type I) (Fig. 3).

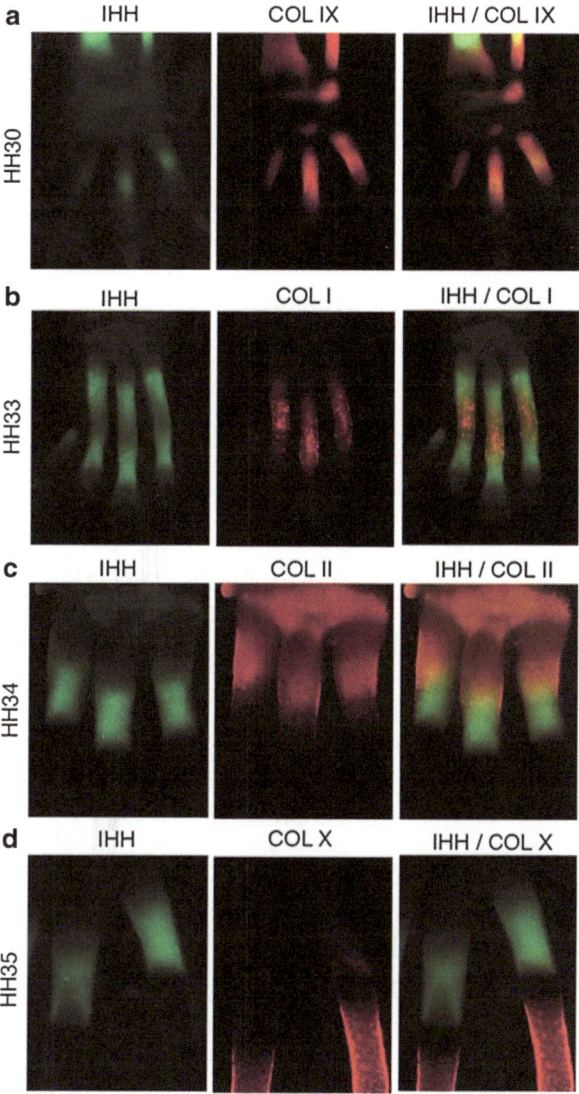

Fig. 3 IHH and collagen expression: IHH expression pattern can be followed until late stages and placed in context using other markers of cartilage differentiation, such as different collagen types, in the developing limbs. (**a**) HH30: Collagen IX expression reveals the forming cartilaginous elements of the hind limb of a chicken embryo, as IHH reveals the maturation of cartilage and differentiation of chondrocytes in the center of some of the elements. (**b**) HH33: Expression of IHH in the metatarsals is subdivided in two domains at the diaphysis. Collagen I expressing cells are present in the place where ossification is starting. (**c**) HH34: In the most proximal end of the metatarsus, the limit between the populations of collagen IX-expressing quiescent chondrocytes and pre-hypertrophic chondrocytes can be visualized and contextualized. (**d**) HH35: The place of transition between pre-hypertrophic chondrocytes, expressing IHH, and hypertrophic chondrocytes, expressing collagen X, can be observed in the diaphysis of a metatarsal

2 Materials

2.1 Equipment

1. Incubator with automatic egg turning.
2. Fertilized chicken eggs.
3. Thermoregulated lab oven or water bath.
4. Orbital shaker.
5. Stereoscopic microscope equipped with epi-fluorescence illumination.
6. Dumont fine forceps numbers #5 and #4 (Fine Science Tools).
7. Curved scissors and tweezers.
8. Scintillation vials of 20 or 7 ml.

2.2 Buffers and Non-buffering Solutions

1. Phosphate buffered saline (PBS) pH 7.4.
2. Methanol 100 %.
3. Dimethyl sulfoxide (DMSO).
4. Hydrogen peroxide (H_2O_2), 30 % (100 volumes).
5. Dent's fix: four parts of methanol, one part of DMSO [25].
6. Dent's bleaching: four parts of methanol, one part of DMSO, and one part of H_2O_2 [25].
7. Triton X-100. Prepare a 20 % stock solution in distillate water (*see* **Note 1**).
8. Urea. Prepare a 4 M solution in water.
9. Glycerol.
10. Normal goat serum (*see* **Note 2**).
11. Hyaluronidase Type I from bovine testes (H3506, Sigma-Aldrich). Dilute 2 mg/ml of hyaluronidase in PBS. Bring to pH 6.0 with citric acid. Prepare 1 ml aliquots and freeze.

2.3 Antibodies

1. Shh-antibody (H-160): sc-9024 from Santa Cruz Biotechnology. Rabbit polyclonal IgG, recognizes Hedgehog family proteins in a broad range of vertebrate species.
2. Anti-collagen type II (II-II6B3), type IX (2C2), anti-collagen type X (X-AC9), and anti-collagen type I (M-38) from Developmental Studies Hybridoma Bank (DSHB) made in mouse.
3. Anti-mouse IgG and anti-rabbit IgG made in goat coupled to fluorophores with different wave emissions.

3 Methods

1. To start development of the embryos, place the eggs in a humidified 37.5 °C (100 °F) incubator and incubate until they reach the desired stage of development. Stage according to

Hamburger and Hamilton [26]. You can use a timer to turn incubator on to obtain intermediate stages precisely (*see* **Note 3**).

2. At the desired day of development, remove eggs from the incubator. Rotate eggs gently around the long axis to ensure that the embryo is floating freely on top of the yolk. Place them in ice for 10 min.

3. Open the egg over the air sac (the larger side) and remove carefully the embryos with the help of a curved tweezers to a petri dish with cold PBS. Leave embryos in cold PBS for 5–10 min to wash the excess of blood.

4. In a stereoscopic microscope select the desired structure of the embryo and remove any excess membrane, most of the skin and muscles carefully with fine forceps.

5. Fix the embryos in Dent's fix in the scintillation vials for 2 h at room temperature. Embryos should become white immediately. The volume of fixing solution must be at least 20 times the volume of embryos (*see* **Note 4**).

6. Dehydrate the embryos with three washes of 100 % methanol for 10 min each.

7. Leave the dehydrated embryos to –80 °C freezer overnight. Embryos could be stored in this condition for at least 6 months (*see* **Note 5**).

8. Take embryos from the freezer and leave them to warm at room temperature for 15 min.

9. Exchange the methanol for Dent's bleaching and leave samples 24 h at room temperature near a cold lamp. A successful bleaching is fundamental to quenching blood autofluorescence and improve clearing process.

10. Wash embryos in PBS three times for 10 min each at room temperature. During the last washing, pre-warm the hyaluronidase solution in the thermoregulated lab oven or water bath to 37 °C.

11. Exchange the PBS for the hyaluronidase solution and leave the embryo at 37 °C for 2 h.

12. Wash the embryo once in PBST (PBS with 1 % Triton X-100) at room temperature.

13. Block the embryos in donkey serum for 1 h in agitation at 4 °C. Meanwhile prepare the primary antibody dilution in PBST/5 % DMSO/2 % goat serum. Anti Shh-antibody (H-160) concentration can be employed between 1:100 and 1:200. Anti-collagens antibody can be employed between 1:10 and 1:50, depending on the batch (*see* **Note 6**).

14. Change blocking solution for primary antibody and leave embryos for 48 h at 4 °C in agitation.

15. Wash embryos six times on PBST for 30 min each. During the last wash, prepare the secondary antibody solution. Dilute both secondary antibodies in PBST/5 % DMSO/2 % Goat serum. We employ the same concentration of secondary antibody employed in standard immunofluorescence assays in sections.

16. Leave embryos overnight in secondary antibody solution at 4 °C.

17. Wash embryos six times on PBST for 30 min each.

18. Prepare a modified version of Scale clearing reagent by mixing 80 % of 4 M urea in water with 20 % glycerol. Do not add Triton X-100. Leave for 5 days at room temperature (*see* **Note 7**).

19. Proper mounting is essential for obtaining good quality images. Photograph in a glass petri dish; 70 % glycerol is an ideal mounting medium for microscopy. Confocal or spinning disk confocal microscopy could be used to obtain more detailed 3D images of the developing cartilages.

4 Notes

1. You can warm the work solution to 60 °C in order to facilitate dilution and avoid excessive bubbling.

2. If your secondary antibody is made in another host species, like horse or donkey, it would be recommended to employ a normal serum from the same species for the preparation of the blocking solution.

3. Incubate eggs with the larger side up and check for 60–70 % humidity.

4. Typical 4 % PFA fixation decreases radically the permeability of the tissue. The primary antibodies here employed recognize their specific epitopes in both formalin and methanol fixed samples. Nevertheless, some antibodies may not work well in methanol fixed samples. In those cases, we recommend to fix in 1 % fresh prepared PFA for a short period, wash out the excess of PFA, and then dehydrate the sample in methanol.

5. The cold methanol treatment is essential for proper fixation and for good quality immunolabeling.

6. Prepare enough solution to cover the sample. In 20 ml scintillation vials, 1 ml is usually enough. In 7 ml scintillation vials 0.5 ml is usually enough.

7. The clearing solution causes swelling of samples. The skeleton is much less affected than other tissues, but excessive exposition could cause its disarticulation. Higher concentrations of

urea and lower concentration of glycerol favor swelling and could be adjusted by each case. Detergent should be avoided because bubbles could be annoying during fluorescent visualization and photography.

Acknowledgment

This work was supported by FONDAP 15090007 (V.P.), Fondecyt grant 1110237 (V.P.), Fondecyt doctoral grant 24100058 (J.B.)

References

1. Bisgard JD, Bisgard ME (1935) Longitudinal growth of long bones. Arch Surg 31(4): 568–578

2. Fell HB (1925) The histogenesis of cartilage and bone in the long bones of the embryonic fowl. J Morphol 40(3):417–459

3. Crombrugghe B, Lefebvre V, Nakashima K (2001) Regulatory mechanisms in the pathways of cartilage and bone formation. Curr Opin Cell Biol 13(6):721–728

4. Maes C, Kobayashi T et al (2010) Osteoblast precursors, but not mature osteoblasts, move into developing and fractured bones along with invading blood vessels. Dev Cell 19(2):329–344

5. Vortkamp A, Lee K et al (1996) Regulation of rate of cartilage differentiation by Indian hedgehog and PTH-related protein. Science 273(5275):613–622

6. Mak K, Chen M et al (2006) Wnt/{beta}-catenin signaling interacts differentially with Ihh signaling in controlling endochondral bone and synovial joint formation. Development 133(18):3695

7. Chung U, Schipani E et al (2001) Indian hedgehog couples chondrogenesis to osteogenesis in endochondral bone development. J Clin Invest 107(3):295–304

8. Pathi S, Rutenberg J et al (1999) Interaction of Ihh and BMP/Noggin signaling during cartilage differentiation. Dev Biol 209(2):239–253

9. Naski MC, Colvin JS et al (1998) Repression of hedgehog signaling and BMP4 expression in growth plate cartilage by fibroblast growth factor receptor 3. Development 125(24): 4977–4988

10. Kronenberg H, Lee K et al (1997) Parathyroid hormone-related protein and Indian hedgehog control the pace of cartilage differentiation. J Endocrinol 154(3 Suppl):S39–S45

11. Minina E, Wenzel H et al (2001) BMP and Ihh/PTHrP signaling interact to coordinate chondrocyte proliferation and differentiation. Development 128(22):4523–4534

12. Koziel L, Wuelling M et al (2005) Gli3 acts as a repressor downstream of Ihh in regulating two distinct steps of chondrocyte differentiation. Development 132(23):5249–5260

13. Karp SJ, Schipani E et al (2000) Indian hedgehog coordinates endochondral bone growth and morphogenesis via parathyroid hormone related-protein-dependent and-independent pathways. Development 127(3): 543–548

14. Long F, Chung U-i et al (2004) Ihh signaling is directly required for the osteoblast lineage in the endochondral skeleton. Development 131(6):1309–1318

15. Hammond CL, Schulte-Merker S (2009) Two populations of endochondral osteoblasts with differential sensitivity to Hedgehog signalling. Development 136(23):3991–4000

16. Zhou J, Meng J et al (2007) IHH and FGF8 coregulate elongation of digit primordia. Biochem Biophys Res Commun 363(3): 513–518

17. Nagashima H, Sugahara F et al (2009) Evolution of the turtle body plan by the folding and creation of new muscle connections. Science 325(5937):193–196

18. Klymkowsky MW, Hanken J (1991) Whole-mount staining of Xenopus and other vertebrates. Methods Cell Biol 36:419–441

19. Hama H, Kurokawa H et al (2011) Scale: a chemical approach for fluorescence imaging and reconstruction of transparent mouse brain. Nat Neurosci 14(11):1481–1488

20. Ke M-T, Fujimoto S, Imai T (2013) SeeDB: a simple and morphology-preserving optical

clearing agent for neuronal circuit reconstruction. Nat Neurosci 16(8):1154–1161

21. Kuwajima T, Sitko AA et al (2013) ClearT: a detergent- and solvent-free clearing method for neuronal and non-neuronal tissue. Development 140(6):1364–1368

22. Johnson RL, Tabin CJ (1997) Molecular models for vertebrate limb development. Cell 90: 979–990

23. Saunders JW (1948) The proximo-distal sequence of origin of the parts of the chick wing and the role of the ectoderm. J Exp Zool 108(3):363–403

24. Sherman A, McGrew M et al (2010) The Roslin Institute Transgenic Chicken Facility: developments in chicken transgenesis. Transgenic Res 19(1):151

25. Dent JA, Klymkowsky MW (1989) Whole-mount analyses of cytoskeletal reorganization and function during oogenesis and early embryogenesis in Xenopus. In: Schatten H, Schatten G (eds) The cell biology of fertilization. Academic, New York, NY, pp 63–103

26. Hamburger V, Hamilton HL (1951) A series of normal stages in the development of the chick embryo. J Morphol 88(1):49–92

Chapter 15

Methods for Detection of Ptc1-Driven LacZ Expression in Adult Mouse Skin

Donna M. Brennan-Crispi, Mỹ G. Mahoney, and Natalia A. Riobo

Abstract

The Ptc-lacZ reporter mice are a highly utilized animal model for studying both normal tissue development and cancer. Identifying cell specific activation of Hedgehog (Hh) signaling is essential to understand the effects of this critical and complex signaling pathway. β-gal detection in tissues can be difficult, with various staining procedures yielding differential results. Thus, detailed information on staining protocols is essential for determining the ideal method for a given study. Furthermore, immunohistochemical staining of X-Gal stained tissues can provide further insight into other key players in Hh signaling activation.

Key words Ptc-lacZ, β-galactosidase (β-gal), X-Gal staining, Histology, Immunohistochemistry

1 Introduction

Reporter genes have proven to be invaluable tools to researchers across the spectrum of biological sciences (for review *see* ref. [1]). One common reporter is the *LacZ* fusion gene. First reported 1980 [2], since then *lacZ* genes have been utilized in a variety of research applications from bacteria colony screening to generation of reporter mice. The lacZ gene encodes the β-galactosidase (β-gal) enzyme, which in E. coli hydrolyzes lactose to galactose and glucose. Scientists have capitalized on the promiscuity of the enzyme and identified several other substrates, including 5-bromo-4-chloro-3-indolyl-β-D-galactopyranoside (X-Gal): which when hydrolyzed by β-gal results in blue product. This colorimetric readout allows for expedient assessment of β-gal reporter activity in many applications.

The earliest use of the *lacZ* reporter in murine models of Hh signaling was reported by the Scott lab with the development of a Ptc1$^{+/lacZ}$ mutant mouse [3]. In this model, regions of exons one and two of Ptc1 were replaced with a functional *lacZ* gene, thus these mice are both heterozygous for Ptc1 and serve as reporters of pathway activation, with β-gal being expressed wherever the

Natalia A. Riobo (ed.), *Hedgehog Signaling Protocols*, Methods in Molecular Biology, vol. 1322,
DOI 10.1007/978-1-4939-2772-2_15, © Springer Science+Business Media New York 2015

canonical pathway is activated. The Scott lab originally character-ized this model for neural tube development and medulloblastoma formation, and it has been used extensively in these fields of research [4–7]. In addition these mice, and variations thereof, have also proven to be highly effective tools in studying the develop-ment of basal cell carcinomas [8–10]. Following in the footsteps of the $Ptc1^{+/lacZ}$ model, additional $lacZ$ reporter mice have been devel-oped to study Hh pathway activity including Shh [11] and Gli1 [12] models. Together these animal models have provided scien-tists with a comprehensive view of Hh pathway players in tissue development and carcinogenesis.

Identification of β-gal expression in these mice is critical to understand temporal and cell type-specific Hh pathway activation in vivo. Here we present a series of protocols designed to detect β-gal expression in the skin of $Ptc1^{+/lacZ}$ mouse model. As Hh pathway activation is also important in hair follicle cycling [13], we will use the hair follicle as a control for our staining procedures. These methods detail two different methods of X-Gal staining in skin, as well as variations to nuclear fast red, histological, and immunohistochemical staining techniques. We note that the pro-tocols here have been tested to various levels in other $Ptc1^{+/lacZ}$ tissues besides skin, and may prove applicable in other lacZ reporter mice as well.

2 Materials

2.1 X-Gal Staining of Frozen Tissue Sections and Nuclear Fast Red Counterstaining

1. Cryostat.

2. Incubator set to 37 °C.

3. Rocker.

4. Slide staining setup: can be series of Coplin jars, staining boxes, Tissue-Tek Slide Staining Set or similar, containing xylenes, ethanols, etc. as needed and container for rocking slides.

5. Fume hood.

6. Curved forceps.

7. Surgical tools.

8. Humidified chamber (see **Note 1**).

9. Basemolds (Fisher Scientific, 22-363-556; or similar).

10. Dry ice (optional).

11. Fisherbrand Colorfrost Plus Microscope Slides (Fisher Scientific, 12-550-17; or similar).

12. Coverslips (Fisher Scientific, 12-548-5D; or similar).

13. Microscope slide pen (StatLab, SMP-BK; or similar) (see **Note 2**).

14. Liquid blocker super pap pen (Newcomer supply, 6505; or similar).

15. OCT Cryopreservation compound (Andwin Scientific, 4583).

16. Formalin, Buffered 10 % (Fisher Scientific, SF99; or similar); prechilled to 4 °C.

17. Phosphate buffered saline (PBS).

18. Distilled water.

19. X-Gal (Fisher Scientific, BP1615-100; or similar) diluted at 40 mg/ml in N,N-dimethylformamide (DMF) (*see* **Note 3**). Dissolve 100 mg X-Gal in 2.5 ml DMF, aliquot into 500 and 50 µl volumes, and store at –20 °C in dark.

20. 0.5 M potassium ferricyanide ($K_3Fe(CN)_6$), store at RT in the dark.

21. 0.5 M potassium ferrocyanide trihydrate ($K_4Fe(CN)_6 \cdot 3H_2O$), store at RT in the dark.

22. 1 M magnesium chloride ($MgCl_2$) (*see* **Note 4**).

23. Nuclear Fast Red (Rowley Biomedical, J-601-3).

24. Glycerol.

25. Nail polish.

26. Ethanol: 100, 95, 75 % v/v.

27. Xylenes (Fisher Scientific, X-4).

28. Permount mounting medium (Fisher Scientific, SP15).

2.2 Whole-Mount X-Gal Staining

1. Rocker.

2. Incubator set to 37 °C.

3. Forceps.

4. Stirring hot plate, set to 60 °C.

5. Surgical tools.

6. 14 ml culture tube with cap.

7. 4 ml culture tube with cap, or other small container with lid.

8. Base molds (Fisher Scientific, 22-363-556; or similar).

9. Transfer pipettes.

10. Paraformaldehyde (PFA).

11. 0.45 µm filter unit.

12. 10 N NaOH.

13. Phosphate buffered saline (PBS).

14. Distilled water.

15. X-Gal (Fisher Scientific, BP1615-100; or similar) diluted at 40 mg/ml in N,N-dimethylformamide (DMF).

16. 0.5 M potassium ferricyanide ($K_3Fe(CN)_6$).

17. 0.5 M potassium ferrocyanide trihydrate ($K_4Fe(CN)_6 \cdot 3H_2O$).

18. 1 M magnesium chloride ($MgCl_2$).

19. Deoxycholate, 5 % stock solution.

20. Nonidet P-40 (NP-40; Tergitol-type NP-40; nonyl phenoxy-polyethoxyethanol), 1 % stock solution.

21. Formalin, Buffered 10 % (Fisher Scientific, SF99; or similar).

22. OCT Cryopreservation compound (Andwin Scientific, 4583).

2.3 H and E Staining of OCT Sections

1. Rocker.

2. Cryostat.

3. Fume hood.

4. Slide staining set up.

5. Curved forceps.

6. Fisherbrand Colorfrost Plus Microscope Slides (Fisher Scientific, 12-550-17; or similar).

7. Coverslips (Fisher Scientific, 12-548-5D; or similar).

8. Tap water.

9. Distilled water.

10. Ethanol: 100, 95,75 %.

11. Xylenes (Fisher Scientific, X-4).

12. Hematoxylin QS (Vector Laboratories, H-3404).

13. Esoin Y (Surgipath Medical, 1600).

14. Rinse solution: 2 % v/v glacial acetic acid in water.

15. Bluing solution: 0.3 % w/v ammonium hydroxide in 70 % ethanol.

16. Permount mounting medium (Fisher Scientific, SP15).

2.4 Nuclear Fast Red Counterstaining of X-Gal Stained, OCT Tissues

1. Rocker.

2. Cryostat.

3. Slide staining set up.

4. Fume hood.

5. Curved forceps.

6. Fisherbrand Colorfrost Plus Microscope Slides (Fisher Scientific, 12-550-17; or similar).

7. Coverslips (Fisher Scientific, 12-548-5D; or similar).

8. Distilled water.

9. Nuclear Fast Red (Rowley Biomedical, J-601-3).

10. Aqueous mounting medium.

11. Ethanol: 100, 95,75, 50 % ethanol.

12. Xylenes (Fisher Scientific, X-4).

13. Hematoxylin QS (Vector Laboratories, H-3404).

14. Esoin Y (Surgipath Medical, 1600).

15. Rinse solution: 2 % glacial acetic acid in water.

16. Bluing solution: 0.3 % ammonium hydroxide in 70 % ethanol.

17. Permount mounting medium (Fisher Scientific, SP15).

2.5 Processing and H and E Staining of Formalin-Fixed, Paraffin-Embedded (FFPE) Sections

1. Tissue-Prep VIP.

2. Embedding Machine.

3. Fume hood.

4. Microtome.

5. Rocker.

6. Cryostat.

7. Slide staining setup.

8. Curved forceps.

9. Incubator/oven heated to 55–65 °C.

10. Tissue pathology cassettes (Fisher Scientific, 22-272-416; or similar).

11. Fisherbrand Colorfrost Plus Microscope Slides (Fisher Scientific, 12-550-17; or similar).

12. Coverslips (Fisher Scientific, 12-548-5D; or similar).

13. Microscope slide pen (StatLab, SMP-BK: or similar).

14. Curved forceps.

15. Formalin (Fisher Scientific, 23-245685).

16. Ethanol for processing (Fisher Scientific, A-405-P4).

17. Xylenes (Fisher Scientific, X-4).

18. Paraplast (Fisher Scientific, T-565).

19. Tap water.

20. Ethanol for staining: 100, 95 %.

21. Hematoxylin (Fisher Scientific, 22-050-111).

22. Clarifier (Fisher Scientific, 22-050-116).

23. Bluing Reagent (Fisher Scientific, 22-050-114).

24. Eosin Y (Fisher Scientific, 22-050-110).

25. Phloxine B: (Fisher Scientific, P-387).

26. Permount mounting medium (Fisher Scientific, SP15).

2.6 Nuclear Fast Red Counterstaining of FFPE X-Gal Stained Tissues

1. Rocker.

2. Microtome.

3. Fume hood.

4. Slide staining set up.

5. Curved forceps.

6. Incubator/Oven heated to 55–65 °C.

7. Fisherbrand Colorfrost Plus Microscope Slides (Fisher Scientific, 12-550-17; or similar).

8. Coverslips (Fisher Scientific, 12-548-5D; or similar).

9. Microscope slide pen (StatLab, SMP-BK; or similar).

10. Glass rod applicator.

11. Xylenes (Fisher Scientific, X-4).

12. Ethanol: 100, 95, 75, 50 %.

13. Distilled water..

14. Nuclear Fast Red (Rowley Biomedical, J-601-3)

15. Aqueous mounting medium.

16. Permount mounting medium (Fisher Scientific, SP15).

2.7 Immunohisto-chemistry on X-Gal Stained Tissues

1. Rocker.

2. Microtome.

3. Fume hood.

4. Slide staining set up.

5. Curved forceps.

6. Incubator heated to 55–65 °C.

7. Refrigerator or incubator set to 4 °C.

8. Inverted light microscope.

9. Microwave (1,050 W).

10. Fisherbrand Colorfrost Plus Microscope Slides (Fisher Scientific, 12-550-17; or similar).

11. Coverslips (Fisher Scientific, 12-548-5D; or similar).

12. Microscope slide pen (StatLab, SMP-BK; or similar).

13. Plastic Coplin jars with lids, small hole punched in lid.

14. 2×1 l Beakers with 500 ml water.

15. Humidified chamber.

16. Liquid blocker super pap pen (Newcomer supply, 6505; or similar).

17. Glass rod applicator.

18. Reagents.

19. Xylenes (Fisher Scientific, X-4).

20. Ethanol: 100, 95, 75, 50 %.

21. Distilled water.

22. Phosphate Buffer Saline (PBS).

23. Vector Antigen Unmasking Solution, 100× stock (Vector Laboratories, H-3300).

24. 3 % hydrogen peroxide solution, store brand is fine.

25. Methanol.

26. Blocking Buffer: 5 % Normal Goat Serum in PBS.

27. Primary Antibodies (of choice).

28. Appropriate EnVision IgG HRP-conjugated secondary antibody for DAB (Dako, Rabbit: K4002, Mouse: K4000).

29. Liquid DAB+: Substrate and Buffer (Dako, K346711-2).

30. Hematoxylin QS (Vector Laboratories, H-3404).

31. Rinse solution: 2 % glacial acetic acid in water.

32. Bluing solution: 0.3 % ammonium hydroxide in 70 % ethanol.

33. Permount mounting medium (Fisher Scientific, SP15).

3   Methods

3.1 X-Gal Staining of Frozen Tissue Sections and Nuclear Fast Red Counterstaining

1. Collect fresh tissue from mice.

2. Use OCT to mount tissue in base mold in correct orientation for cutting.

3. Freeze on dry ice or in –80 °C freezer.

4. Store at –80 °C until ready to section.

5. Section tissues embedded in OCT at 6–12 μm onto Colorfrost Plus slides labeled with HistoPrep pen (*see* **Notes 5** and **6**).

6. Incubate slides flat at RT for 10–20 min to allow tissue to affix to the slide.

7. Store slides at –80 °C until ready to use.

8. Prepare X-Gal Dilution Buffer for tissue section staining and store in the dark at 4 °C:

   (a) 500 μl of 0.5 M potassium ferricyanide Crystalline (final concentration 5 mM).

   (b) 500 μl of 0.5 M potassium ferrocyanide trihydrate (final concentration 5 mM).

   (c) 100 μl of 1 M magnesium chloride (final concentration 2 mM).

   (d) 48.9 ml PBS.

9. Prepare aqueous mounting medium: combine PBS and glycerol in a 1:1 ratio (50 % PBS, 50 % Glycerol), mix well, store in the dark at 4 °C.

10. Prewarm X-Gal dilution buffer (*see* **Note 7**).

11. Equilibrate OCT slides to room temperature, wipe away any excess moisture.

12. Draw box around tissue with pap-pen, allow to dry briefly.

13. Fix slides in 10 % formalin at 4 °C for 10 min.

14. In the meantime, prepare 1 mg/ml X-Gal in dilution buffer: add X-Gal from 40 mg/ml stock to pre-warmed dilution buffer and keep at 37 °C, in dark until ready to use.

15. Wash slides with PBS 3× for 5 min on rocker (*see* **Note 8**).

16. Rinse briefly in distilled water.

17. Blot dry slides and remove excess liquid from tissue, be careful not to damage or dry out tissue.

18. Place slides in humidified chamber.

19. Add X-Gal solution to tissue sections.

20. Incubate O/N at 37 °C, in dark (*see* **Note 9**).

21. Dump off X-Gal solution.

22. Wash slides in PBS 2× for 5 min on rocker.

23. Rinse in distilled water.

24. Add Nuclear Fast Red (Rowley Biomedical, J-601-3) for 3–5 min (*see* **Notes 10** and **11**).

25. Wash in running distilled water.

26. Determine which type of mounting is best for your application (*see* **Note 12**). Proceed to mount slides in aqueous (**step 27**) or Permount medium (**step 28**).

27. For aqueous mounting blot excess water from slide and tissue sections and using a transfer pipette add a thin line of aqueous mounting medium to slide. Gently drop coverslip on top of slide, allow to settle, and remove air bubbles by applying gentle pressure to the coverslip with a pair of curved forceps. Any mounting medium that leaks should be gently blotted away. Seal with nail polish, keep slides flat until nail polish has hardened and place in slide box for long-term storage at 4 °C.

28. For Permount mounting (Fig. 1a), first incubate slides 3–5 min each in 75 % ethanol, 95 % ethanol and 100 % ethanol, blotting between each change in solution, then incubate slides 10–15 min in xylenes. Let air dry 10–15 min and place one drop of Permount onto each tissue section using glass rod applicator. Flip slide over onto coverslip, applying even, gentle pressure. Mounting medium can be spread over tissue and air bubbles removed by applying pressure to the coverslip with a pair of curved forceps. Remove any mounting medium that leaks by dipping slides in xylenes and gently wiping, if necessary. Slides should remain flat for approximately 24 h until Permount is set; afterwards slides can be stored in slide box.

3.2 Whole-Mount X-Gal Staining

In addition to performing X-Gal staining in tissue sections, it is possible and many times convenient to stain with X-Gal the whole tissue, followed by sectioning and counterstaining. This procedure involves fixation of the whole tissue in PFA, X-Gal staining, and a second fixation step that varies according to the embedding method (OCT or paraffin). Examples of sections that were whole-mount stained are shown in Figs. 1b, 2, and 3.

1. Determine hotplate setting to heat 400 ml of water in 2 l beaker to ~60 °C, bring water to temperature.

2. To prepare 4 % paraformaldehyde (PFA), 1 l, ~pH 7.3, combine in a 1 l bottle: 980 ml RNAse free PBS, 100 µl 10 N NaOH, and 40 g paraformaldehyde (*see* **Note 13**).

3. Place bottle in beaker of heated water at ~60 °C and monitor temperature while stirring the solution; the fixative solution should become clear when it is heated to 62–64 °C.

4. Once solution is clear, remove beaker from hot plate and allow to cool to room temperature.

5. Filter fixative using 0.45 µm filter, aliquot into small volumes, and store at –20 °C, in dark.

6. Prepare 50 ml of X-Gal Dilution Buffer for whole-mount staining. Combine the following and store dilution buffer in dark at 4 °C:

    (a) 500 µl of 0.5 M potassium ferricyanide crystalline (final concentration 5 mM).

    (b) 500 µl of 0.5 M potassium ferrocyanide trihydrate (final concentration 5 mM).

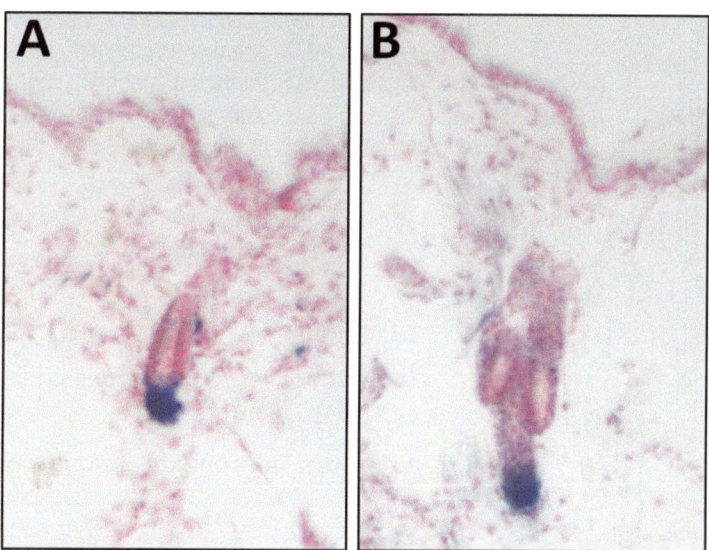

Fig. 1 Comparison of section vs. whole-mount staining. (**a**) 12 µm OCT tissue sections was stained with X-Gal as indicated above and mounted in Permount. (**b**) Whole-mount stained tissues were embedded in OCT, sectioned at 12 µm, and counterstained with Nuclear Fast Red as indicated above. Note the staining intensity is similar in these two hair follicles, but nuclear counterstain is more discrete in whole-mount stained tissues

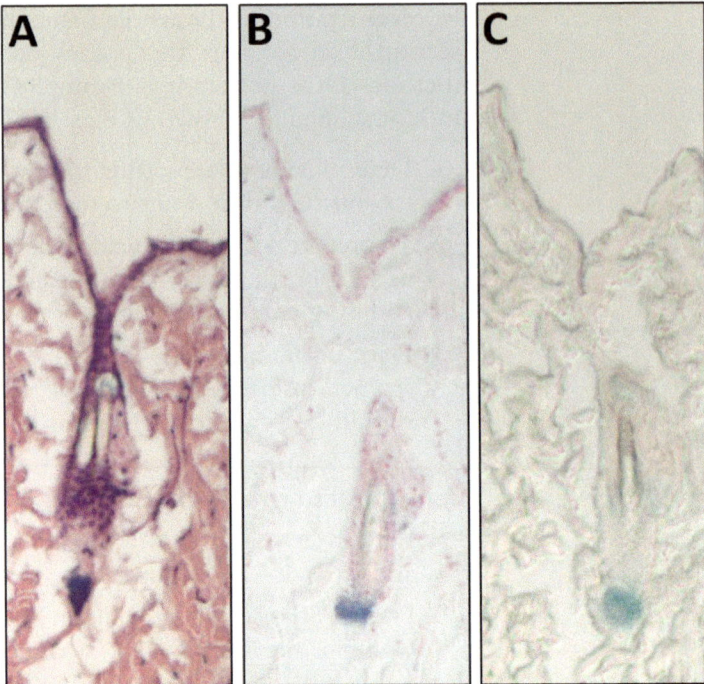

Fig. 2 Whole-mount staining variations on OCT sections. Skin was whole-mount stained, embedded in OCT, sectioned to 6 μm, and stained as indicated. (**a**) Tissue was counterstained with H and E, X-Gal staining can be difficult to see with this method, but there is a color difference between X-Gal (*blue*) and hematoxylin (*purple*). Note the tissue morphology is well defined due to H and E staining. (**b**, **c**) Sections were counterstained with Nuclear Fast Red and mounted using either Permount (**b**) or aqueous mounting medium (**c**). Tissue morphology is better visualized with Permount mounting, but aqueous mounting results in the more classic blue color associated with X-Gal staining. Note that these are the sequential sections of the same follicle

    (c) 100 μl of 1 M magnesium chloride (final concentration 2 mM).

    (d) 500 μl of 5 % deoxycholate (final concentration 0.05 %).

    (e) 1 ml of 1 % NP-40 (final concentration 0.02 %).

    (f) 47.4 ml PBS.

7. Pre-warm appropriate amount of X-Gal dilution buffer to 37 °C.

8. Collect fresh tissue from mice.

9. Fix tissues in 4 % PFA for 20–40 min at room temperature in 14 ml culture tubes (*see* **Notes 14** and **15**).

10. Remove PFA with transfer pipette and rinse once with 5 ml PBS.

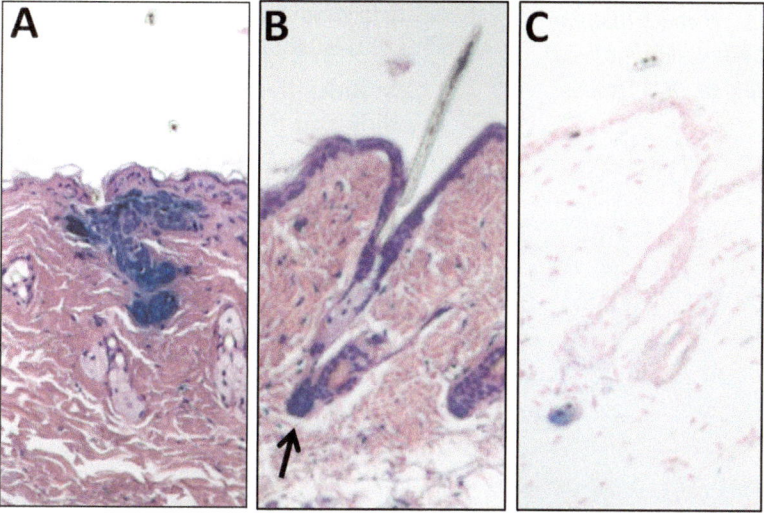

Fig. 3 Whole-mount staining variations on FFPE sections. Skin was whole-mount stained, processed for FFPE, sectioned to 5 μm, and stained as indicated. Tissue was stained with H and E. Basal cell carcinomas stain strongly with X-Gal (**a**). Note that very weak staining is not apparent via this method, though there is a slight *blue* color evident in the dermal papilla of the hair follicle (**b**, *arrow*). Nuclear Fast Red counterstaining with Permount mounting (**c**) yields a very defined tissue morphology. At standard histological sectioning thickness of 5 μm, low-level staining is visible, but not as strong as other methods. Thicker sections can enhance X-Gal signal, but will result in less defined morphology. A balance between signal strength and tissue morphology must be determined based on tissue and signal strength

11. Wash twice with 10 ml PBS for 10 min on rocker.

12. Prepare 1 mg/ml X-Gal in dilution buffer: dilute X-Gal stock 1:40 in pre-warmed dilution buffer and keep at 37 °C, in dark until ready (*see* **Note 7**).

13. Remove tissue from PBS and blot gently on paper towel before adding to tube with X-Gal solution. Use enough X-Gal buffer to 4 ml culture tube to completely cover tissue.

14. Incubate O/N at 37 °C in dark.

15. Remove tissue, dip in PBS to rinse, blot gently, and place in 14 ml culture tube with 10 ml PBS (*see* **Note 16**).

16. Wash three times with 10 ml PBS for 10 min on rocker.

17. Post Fix/Embed using method of choice (*see* **Note 17**). For formalin-fixed, paraffin-embedded (FFPE) tissues, fix overnight in 10 % Formalin. For frozen tissues, use OCT to mount tissue in base mold in correct orientation for cutting, and freeze on dry ice or in −80 °C freezer and store at −80 °C.

3.3 H and E Staining of Whole-Mount X-Gal Stained, OCT Tissues

1. Section tissues embedded in OCT at 6–12 μm onto Colorfrost Plus slides labeled with HistoPrep pen.

2. Leave slides flat at RT for 5–20 min to allow tissue to affix to the slide and then store at –80 °C until ready to use.

3. Briefly equilibrate OCT slides to room temperature, wipe away any excess moisture.

4. Fix slides by dipping in 95 % ethanol ten times

5. Rinse in running tap water for 1–2 min.

6. Rinse in running distilled water for 1–2 min.

7. Blot slides and remove excess liquid.

8. Add hematoxylin QS directly to tissue and stain for 20–30 s

9. Dump of excess hematoxylin and place in distilled water.

10. Rinse in running distilled water until clear; time depends on size of container.

11. Blot slides on paper towel.

12. Dip ten times in rinse solution, blot on paper towel.

13. Incubate 1 min in bluing solution, blot on paper towel.

14. Dip slides ten times in water to rinse, blot on paper towel.

15. Dip slides ten times in 95 % ethanol, blot on paper towel.

16. Dip slides 1–2 (quick) in Eosin, blot on paper towel.

17. Dip slides ten times in 95 % ethanol, blot on paper towel.

18. Dip slides ten times in 95 % ethanol, blot on paper towel.

19. Mount slides using Permount as indicated in Subheading 3.1. An example is shown in Fig. 2a.

3.4 Nuclear Fast Red Counterstaining of Whole-Mount X-Gal Stained, OCT Tissues

1. Prepare slides as for H and E staining, see detailed directions above.

2. Equilibrate OCT slides to room temperature.

3. Wash twice in PBS for 5 min.

4. Wash briefly (1–3 min) in distilled water.

5. Incubate 3–5 min in Nuclear Fast Red (*see* **Notes 10** and **11**).

6. Wash 1–3 min in distilled water.

7. Mount slides using Permount (Fig. 2b) or aqueous mounting medium (Fig. 2c), as indicated in Subheading 3.1.

3.5 H and E Staining of X-Gal Stained, FFPE Tissues

1. Prepare 1 % phloxine B in Esoin.

2. Fix tissues with formalin in tissue pathology cassettes and process on a Tissue-Prep VIP tissue processor as follows (*see* **Notes 18** and **19**):

    (a) Formalin—10 min.

    (b) 70 % ethanol—30 min.

(c) 95 % ethanol—50 min.

(d) 95 % ethanol—60 min.

(e) 100 % ethanol—50 min.

(f) 100 % ethanol—60 min

(g) 50 % ethanol/50 % xylenes—30 min.

(h) Xylenes—45 min.

(i) Xylenes—60 min.

(j) Xylenes—60 min.

(k) Paraffin—30 min.

(l) Paraffin—90 min.

(m) Paraffin—120 min.

3. Embed tissues in paraffin blocks.

4. Section blocks at 5–12 μm onto Colorfrost Plus slides labeled with HistoPrep pen.

5. Bake slides at 55–65 °C for 10–15 min to allow to adhere to slide (*see* **Note 20**).

6. Store at room temperature until ready to use.

7. Bake slides at 55–65 °C for 30–60 min to melt paraffin.

8. Soak slides in xylene for 1–3 min, three times.

9. Soak slides in 100 % ethanol for 30–60 s (5–7 dips), two times.

10. Soak slides in 95 % ethanol for 30–60 s (5–7 dips).

11. Wash with running tap water 1–2 min.

12. Soak slides in Hematoxylin for 1 min (if the reagent is old, use 3 min).

13. Wash with running tap water 1–2 min, until water runs clear.

14. Soak slides in Clarifier 1 solution for 2 quick dips.

15. Wash with running tap water 1–2 min.

16. Soak slides in Bluing Reagent for 15 s (5–7 dips).

17. Wash with running tap water 1–2 min.

18. Soak slides in 95 % ethanol for 15–30 s.

19. Soak slides in Eosin for 3 s (if the reagent is old, use 8–9 s).

20. Quick dip slides 5–8 times in 95 % ethanol, repeat two times in fresh 95 % ethanol.

21. Quick dip slides 5–8 times in 100 % ethanol, repeat two times in fresh 100 % ethanol.

22. Soak slides in xylene 1 min for four times (can go longer).

23. Mount slides using Permount as indicated in Subheading 3.1 (Fig. 3a).

3.6 Nuclear Fast Red Counterstaining of X-Gal Stained, FFPE Tissues

1. Tissue is processed and slides are prepared as indicated in Subheading 3.5.

2. Bake slides at 55–65 °C for 30–60 min (*see* **Note 20**).

3. Deparaffinize/Rehydrate slides—Incubate as follows, blotting in between each solution:

    (a) Xylenes: 10 min 2×, 5 min 1×.

    (b) 100 % ethanol: Rinse, 5 min 2×.

    (c) 95 % ethanol: 5 min 2×.

    (d) 75 % ethanol 3–5 min 1×.

    (e) 50 % ethanol 5 min 1×.

    (f) Distilled, deionized water 1–2 min.

4. Counterstain with Nuclear Fast Red for 3–5 min (*see* **Notes 10** and **11**).

5. Wash 1–3 min in distilled water.

6. Mount slides using Permount as indicated in Subheading 3.1 (Fig. 3b).

3.7 Immunohisto-chemistry on X-Gal Stained, FFPE Tissues

IHC has successfully been performed using several antibodies with this protocol. While citrate-based antigen retrieval methods are covered here, other methods such as EDTA or Tris-Tween antigen retrievals may also be effective. Please note that optimization for individual antibodies will be required. Here anti-mouse involucrin (BioLegend; Dedham, MA) at 1:2,000 and anti-Flag pAb (Sigma Aldrich; St. Louis, MO) at 1:1,000 were used effectively.

1. Tissues are sectioned a 5–12 μm onto Colorfrost Plus slides labeled with HistoPrep pen.

2. Bake slides at 55–65 °C for 10–15 min to allow to adhere to slide (*see* **Note 20**).

3. Store at room temperature until ready to use.

4. Prepare blocking buffer: 5 % normal goat serum in PBS.

5. Bake slides at 55–65 °C for 30–60 min.

6. Deparaffinize/Rehydrate slides as indicated above.

7. While slides are going through xylenes and ethanols, prepare 1× Antigen Unmasking Solution from 100× stock, enough to fill plastic Coplin jar so that slides are completely covered.

8. Prepare peroxidase blocking solution: 0.3 % hydrogen peroxide (H_2O_2) in methanol, enough to fill plastic Coplin jar so that tissues are completely covered.

9. Prefill 2 × 1 l beakers with 500 ml tap water; when slides are in water, preheat beakers in microwave for 1 min.

10. Remove slides from water, blot and wipe away excess liquid.

11. Place slides in Coplin jar with Antigen Unmasking Solution.

12. Place Coplin jar(s) in beaker(s) of water (*see* **Note 21**).

13. Microwave for 5 min on high (*see* **Note 22**).

14. Remove and let cool on bench for 20–45 min.

15. Remove slides, and blot excess liquid.

16. Draw box around tissue with pap-pen, allow to dry briefly.

17. Wash slides in PBS for 5 min.

18. Remove slides, and blot excess liquid; be careful not to let slides over dry.

19. Block 20 min in 0.3 % H_2O_2 in methanol.

20. Wash slides in PBS for 5 min.

21. Remove slides, and blot excess liquid; be careful not to let slides over dry.

22. Block at RT in 5 % normal goat serum/PBS for 1 h.

23. Incubate slides with primary antibody O/N at 4 °C in blocking solution. Dilution of the antibody is variable and needs to be determined (example anti-involucrin 1:2,000).

24. Wash twice with PBS for 10 min each on rocker.

25. Incubate with EnVision IgG HRP-conjugated secondary antibody for 1 h at RT.

26. Wash in PBS 3× at 10 min on rocker.

27. Prepare fresh DAB substrate solution.

28. Incubate 1–30 min in fresh DAB, length of time determined by antibody (for anti-involucrin, 7.5 min).

29. Stop reaction via immersion in water.

30. Wash slides in distilled water for 3–5 min.

31. Add hematoxylin QS directly to tissue and stain for 20–30 s.

32. Dump off excess hematoxylin and place in distilled water.

33. Rinse in running distilled water until clear, time depends on size of container.

34. Blot slides on paper towel.

35. Dip ten times in rinse solution, blot on paper towel.

36. Incubate 1 min in bluing solution, blot on paper towel.

37. Dip slides ten times in water to rinse, blot on paper towel.

38. Mount with Permount, see detailed directions in Subheading 3.1. An example of IHC using anti-involucrin Ab and X-gal in mouse basal cell carcinoma is shown in Fig. 4.

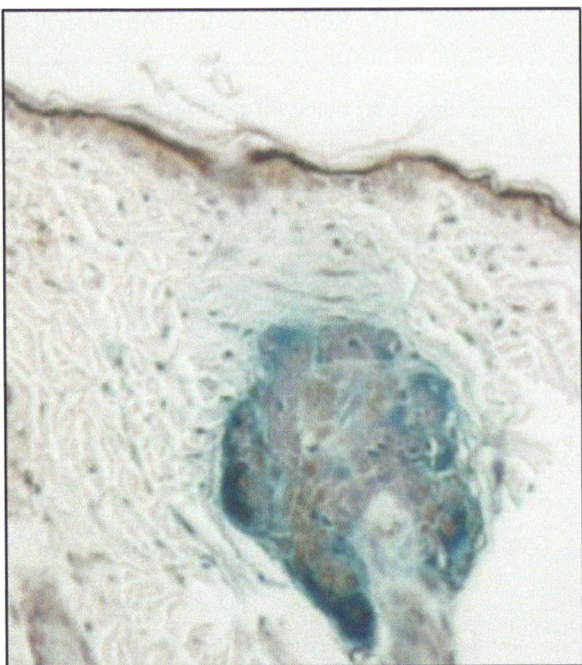

Fig. 4 IHC on whole-mount stained, FFPE tissues. Skin with basal cell carcinomas were whole-mount stained, processed for FFPE, sectioned to 5 μm, and subjected to IHC as indicated. Anti-involucrin antibodies were used to label the differentiated epidermis, as expected staining is observed in the differentiated layer of the epidermis. X-Gal staining is clearly visible in a BCC lesion emanating from a hair follicle

4 Notes

1. Humidified chambers are simply containers with lids, in which the bottom has been lined with moistened paper towels or filter paper. These are essential to prevent tissues from drying out.

2. Microscope pens (StatMark, HistoPrep, etc.) are specialized pens resistant to solvents (xylenes, ethanol, etc.) and must be used to label slides, sharpie markers will not work. Ensure ink has time to dry before immersing slides in any liquid including aqueous buffers as the ink may run otherwise. Labeling the day before or after mounting is highly recommended.

3. Alternatively, Promega X-Gal (V394) which is 50 mg/ml in DMF (N,N-Dimethylformamide) has been tested in some staining conditions mentioned here. It is diluted to 1:50 in X-Gal dilution buffer for a working concentration of 1 mg/ml.

4. Remember when preparing reagents, that water weight is taken into account with calculations, particularly with magnesium chloride.

5. When sectioning OCT tissues and preparing slides, it is important to (1) allow slides to equilibrate to the sectioning temperature of the cryostat (–18 to –25 °C) so that tissues are not too brittle for sectioning, and (2) allow adequate time at room temperature for sections to adhere to the slide. If not, tissue is prone to being damaged and washed away with washes.

6. Tissue sections are provided in a range that has proven effective in several tissues tested. Figure legends include detailed information on skin section thickness used to obtain each image. Note that for some conditions a range in tissue thickness will yield similar results. Optimal section thickness for X-Gal staining and IHC staining should be optimized on a case by case basis. In general, X-Gal staining increases in intensity with thickness, but individual cell morphology may be lost. Furthermore, and particularly with FFPE samples, the thicker the section the more intense the H and E staining can be, and thus it can be more difficult to see the β-gal signal. Shortening time in hematoxylin and eosin may prove useful in thicker sections.

7. X-Gal dilution buffer must be pre-warmed to 37 °C to prevent X-Gal from precipitating out of solution. Be sure to allow adequate time for buffer to get to temperature.

8. Excess liquid must be removed quickly and carefully to avoid tissue damage. It is imperative that tissues not be allowed to dry out.

9. Note that since tissues are incubated overnight at 37 °C in X-Gal, the solution is more prone to evaporation that if incubation were carried out at RT or 4 °C; therefore, enough solution must be used to ensure that solution does not completely evaporate during incubation.

10. Nuclear Fast Red may develop a precipitate over time. If it does, it needs to be filtered prior to use, a coffee filter works well.

11. Sections can either be covered dropwise with Nuclear Fast Red, or the entire slide can be submerged. Furthermore, washing post Nuclear Fast Red staining is dependent on the method. If the tissue sections are simply covered, slides can be rinsed with distilled water in slide box on rocker for several minutes. If the slides were submerged, they should be transferred to a new container with distilled water and the rinsed with running distilled water until clear.

12. The type of mounting depends upon desired staining. Aqueous mounting yields lighter, less "nuclear-localized" Nuclear Fast Red staining; and while it is quick and easy, slides must be stored at 4 °C and are not permanent. With very careful storage, slides may last weeks to months, but it is recommended that any images are taken promptly as the counterstain may fade over time. Permount mounting gives stronger, nuclear

localized staining, and slides may be stored indefinitely at room temperature. Weak X-Gal signal can still be seen, though the contrast with increased counterstain may not be ideal in all cases. In addition, ethanol-xylene preparation will remove all traces of the pap pen, so if multiple sections are on a slide, be sure to be able to tell them apart. Please *see* Fig. 1 and choose mounting method according to specific requirements.

13. 4 % PFA is prepared in bulk and kept at –20 °C for long-term storage. Determining pH of this solution is not necessary when following this protocol. PFA should be frozen in appropriate aliquots, as it cannot be re-frozen and must be used within 1 week of thawing.

14. Remember to allow adequate time to thaw PFA and to ensure all PDA is in solution. Incubating in a 37 °C water bath and vortexing can speed up the process. Thawing O/N at 4 °C is also typically effective.

15. Be careful not to over-fix samples. It is better to begin washes and let tissues sit in PBS rather than extended time in PFA as β-gal activity may be lost.

16. Tissues harboring the β-gal reporter should show grossly visible signs of blue signal. Note that the localization and intensity will vary based on Hh activity. In addition, a nonspecific light blue color may be present on some tissues, which is sometimes reduced after washing. Regardless, it is important to analyze tissue sections to be certain of signal and localization and intensity of expression.

17. Whole-mount tissues are sent out for processing, paraffin-embedding, and histology staining. The protocol here is that used by the dermatology core facility.

18. The reagents for the staining set up such as xylenes, ethanol dilutions, and staining reagents may often be used for 1–2 weeks depending upon frequency of use. Xylenes and eosin should be used under a chemical fume hood. The xylenes and ethanol dilutions may be used in both deparaffinization and dehydration before Permount mounting, but it is best to use the "cleanest" solution (e.g., do not use the first xylene that is used for deparaffinization as there will be more paraffin in the solution.)

19. There is some flexibility with the ethanol/xylene dehydration steps. The minimum times are listed, but the entire process can be spread out over several hours if need be. Ethanol 15–30 min, and xylenes up to 2 h if necessary.

20. FFPE sections should be baked to the slides to ensure the tissues adhere well. Note that the temperature indicated depends upon the melting temperature of the wax in which the tissues were embedded.

21. Coplin jars should be surrounded by water, but not submerged. This method allows for more even heating and less evaporation of antigen unmasking solution.

22. Heating time is based on a 1,050 W microwave, though 4.5 min in a 1,500 W microwave also proved effective for several antibodies. Again, timing will need to be optimized based on antibody. Additionally, cooling time may also vary depending upon antibodies.

References

1. Ghim CM, Lee SK, Takayama S, Mitchell R (2010) The art of reporter proteins in science: past, present and future applications. BMB Rep 43:451–460

2. Casadaban MJ, Chou J, Cohen SN (1980) In vitro gene fusions that join an enzymatically active beta-galactosidase segment to amino-terminal fragments of exogenous proteins: Escherichia coli plasmid vectors for the detection and cloning of translational initiation signals. J Bacteriol 143:971–980

3. Goodrich LV, Milenkovic L, Higgins KM, Scott MP (1997) Altered neural cell fates and medulloblastoma in mouse patched mutants. Science 277:1109–1113

4. Zurawel RH, Allen C, Wechsler-Reya R, Scott MP, Raffel C (2000) Evidence that haploinsufficiency of Ptch leads to medulloblastoma in mice. Genes Chromosomes Cancer 28:77–81

5. Wetmore C, Eberhart DE, Curran T (2000) The normal patched allele is expressed in medulloblastomas from mice with heterozygous germ-line mutation of patched. Cancer Res 60:2239–2246

6. Oliver TG, Read TA, Kessler JD, Mehmeti A, Wells JF, Huynh TT et al (2005) Loss of patched and disruption of granule cell development in a pre-neoplastic stage of medulloblastoma. Development 132:2425–2439

7. Mille F, Tamayo-Orrego L, Lévesque M, Remke M, Korshunov A, Cardin J et al (2014) The Shh receptor boc promotes progression of early medulloblastoma to advanced tumors. Dev Cell 31:34–47

8. So PL, Langston AW, Daniallinia N, Hebert JL, Fujimoto MA, Khaimskiy Y et al (2006) Long-term establishment, characterization and manipulation of cell lines from mouse basal cell carcinoma tumors. Exp Dermatol 15:742–750

9. Davari P, Hebert JL, Albertson DG, Huey B, Roy R, Mancianti ML et al (2010) Loss of Blm enhances basal cell carcinoma and rhabdomyosarcoma tumorigenesis in Ptch1+/- mice. Carcinogenesis 31:968–973

10. Wang GY, Wang J, Mancianti ML, Epstein EH Jr (2011) Basal cell carcinomas arise from hair follicle stem cells in Ptch1(+/-) mice. Cancer Cell 19:114–124

11. Jeong J, Mao J, Tenzen T, Kottmann AH, McMahon AP (2004) Hedgehog signaling in the neural crest cells regulates the patterning and growth of facial primordia. Genes Dev 18:937–951

12. Bai CB, Auerbach W, Lee JS, Stephen D, Joyner AL (2002) Gli2, but not Gli1, is required for initial Shh signaling and ectopic activation of the Shh pathway. Development 129:4753–4761

13. Oro AE, Higgins K (2003) Hair cycle regulation of Hedgehog signal reception. Dev Biol 255:238–248

Chapter 16

Determination and Analysis of Cellular Metabolic Changes by Noncanonical Hedgehog Signaling

Raffaele Teperino and John Andrew Pospisilik

Abstract

Hedgehog is a morphogen essential for body patterning and proper embryonic development from flies to humans. Thought quiescent in adults, its inappropriate reactivation is associated with many disparate genetic and sporadic types of human cancers. Recent findings have demonstrated a key, yet unexpected, role of the Hedgehog signaling pathway in metabolic control. Here, we describe a panel of methods to determine and analyze cellular and organismal metabolic changes downstream of the Hedgehog signaling pathway.

Key words Hedgehog, Metabolism, Glucose uptake, Calcium, Ampk

1 Introduction

Obesity and diabetes have reached epidemic proportions with almost one billion people affected worldwide (Diabetes Atlas 6th ed.). This disease burden and the lack of adequate long-term treatments make obesity and diabetes a key health-care challenge of our day and force research to identify novel molecular targets.

Hedgehog (Hh) is an evolutionary conserved morphogen essential for proper embryonic development [1]. The Hedgehog signal transduction pathway is paradigmatic of how a single molecular cue can be translated into an array of different cellular responses depending on the context [2]. A detailed depiction of the pathway goes beyond the scope of this chapter and has been already extensively described [3–5]. It is essential, though, to point out that different modes of Hh signaling exist and are most commonly classified as either canonical or noncanonical. Briefly, canonical signaling involves Hh-dependent activation of the Gli zinc finger transcription factors [6]. Noncanonical Hh signaling, instead, currently refers to receptor-dependent signaling cascades that don't act via the canonical Hh-to-Gli route [7]. These cascades have been classified into two groups: *Type I* signals, which stem from the

Natalia A. Riobo (ed.), *Hedgehog Signaling Protocols*, Methods in Molecular Biology, vol. 1322,
DOI 10.1007/978-1-4939-2772-2_16, © Springer Science+Business Media New York 2015

ligand-binding receptor Patched (Ptch); and *Type II* signals, that stem from the G-protein coupled receptor Smoothened (Smo). A novel *Type II* noncanonical Hh signaling pathway has been recently identified, which rewires cellular metabolism and affects whole body energy homeostasis by inducing insulin-independent glucose uptake in skeletal muscle and brown adipose tissue [8]. Beyond its role in development, inappropriate reactivation of the Hh pathway in adults is associated with multiple human cancers and several pharmaceutical companies have developed specific inhibitors either approved or in advanced phase of clinical development [9, 10]. Interestingly, many of the canonical inhibitors, when used as tool compounds to inhibit the Hh-dependent metabolic response, behave as partial agonists. Thus, the noncanonical metabolic arm of the Hh pathway is both mechanistically and pharmacologically uncoupled from the oncogenic canonical cascade [8]. These findings have tremendous and immediate clinical impact.

Mechanistically, ligand binding to Smo induces Ca^{2+}-dependent activation of the AMP-dependent protein kinase (Ampk), which stimulates the translocation to the plasma membrane of the insulin-responsive glucose transporter Glut4, robust glucose uptake, and a switch of the cellular metabolism away from mitochondrial respiration towards aerobic glycolysis (Fig. 1).

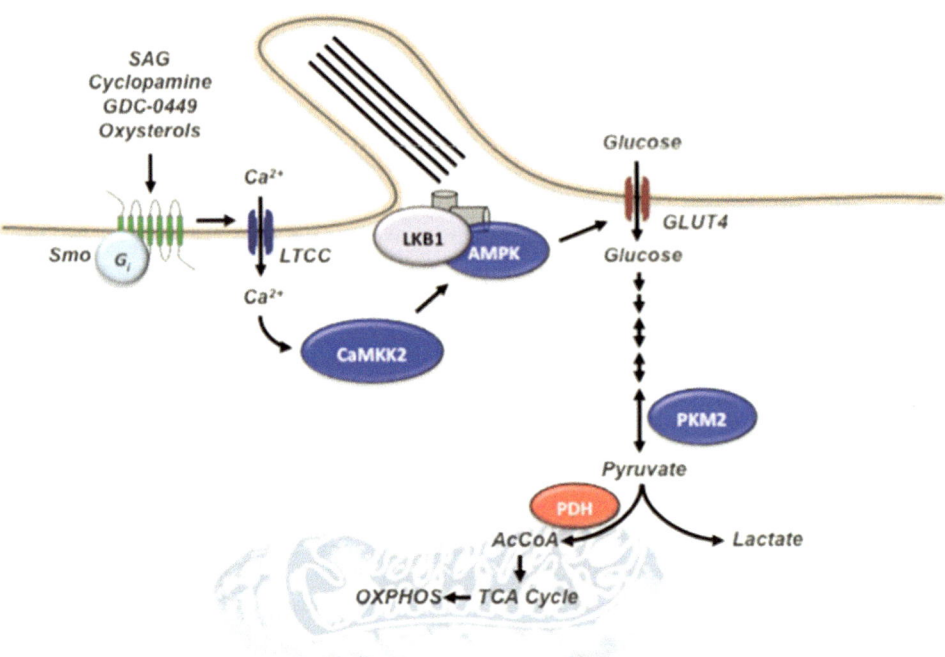

Fig. 1 Smo-Ca^{2+}-Ampk noncanonical hedgehog signaling

The aim of this chapter is to highlight the methods used to characterize this novel Smo-Ampk axis and analyze metabolic consequences in vitro and in vivo. Since many of these technologies have been already established, validated, and widely used, we focus on how they have been adapted to answer our specific questions.

2 Biological Materials

2.1 Cell Lines and Animals

1. Murine 3T3-L1 adipocytes, differentiated as indicated in Subheading 3.1.

2. Murine C2C12 myoblasts.

3. Murine brown adipocytes.

4. Primary human myoblasts.

5. C57BL/6J mice. The mouse strain used to study Hh-dependent metabolic effects is the C57BL/6J, commonly used in metabolic studies for its susceptibility to diet-induced obesity, type-2 diabetes, and atherosclerosis [11].

3 Methods

3.1 Cell Culture

3.1.1 Culture and Differentiation of 3T3-L1 Adipocytes

3T3-L1 cells are the most commonly used cell line to study adipocyte biology. Derived from the 3T3 mouse fibroblasts, they differentiate into functional adipocytes under specific culture conditions.

1. Low-passage cells (max 15) are maintained in Dulbecco's Modified Eagle Medium (DMEM) in the presence of 10 % heat-inactivated calf serum (CS) at 37 °C and 5 % CO_2 (*see* **Note 1**).

2. As the starting point for the differentiation protocol, allow cells to reach the confluence (d_{-2}).

3. Two days post-confluence (d_0), induce differentiation by treating cells with 0.25 mM 3-isobutyl-1-methylxanthine (IBMX), 1 µM dexamethasone, 1.74 µM insulin, and 5 µM troglitazone (TGZ) in DMEM containing 10 % heat-inactivated fetal bovine serum (FBS).

4. 48 h after induction (d_2), replace the medium with DMEM supplemented with 10 % FBS, 1.74 µM insulin, and 5 µM TGZ for additional 48 h (d_4).

5. From d_4 onwards, maintain cells in DMEM supplemented with 10 % FBS and change medium every 48 h (*see* **Note 2**).

6. Before treatments, generally at d_7 of differentiation (*see* **Note 3**), starve mature adipocytes from serum (by changing DMEM 10 % FBS with DMEM supplemented with 1 % bovine serum albumin—BSA) for 16 h. Treatments are done as previously specified (*see* **Note 4**) [8].

3.1.2 Culture and Differentiation of Murine C2C12 Myoblasts

1. Grow low passage myoblasts (max 10) in DMEM supplemented with 10 % FBS (*see* **Note 5**).

2. For differentiation, allow cells to reach 80 % confluence and switch to the differentiation medium consisting of DMEM supplemented with 2 % heat inactivated Horse Serum (HS).

3. Replace medium every 48 h. Cells are considered myotubes after 96 h in differentiation medium, as previously reported [12].

4. As for 3T3-L1 adipocytes, serum-starve C2C12 myotubes for 16 h before any treatment.

3.1.3 Culture and Differentiation of Murine Brown Adipocytes

1. Isolate primary brown pre-adipocytes from interscapular brown adipose tissue depots and differentiate them ex-vivo as previously reported [13].

2. Serum-starve cells for 16 h before any treatment.

3.1.4 Primary Human Myoblasts

1. Follow the procedure published by Gaster et al. [14, 15] to establish cultures of human primary myoblasts and myocytes.

2. Stimulate human primary cells as the other cell culture models.

3.2 Analysis of Cellular Metabolism

3.2.1 Glucose Uptake

Glucose is taken up from the extracellular milieu either passively, via facilitative glucose transporters (GLUT), or actively via kidney-specific symporters, low (SGLT2) or high (SGLT1) affinity transporters that depend on the creation of a Na^+ gradient generated by the activity of the Na^+/K^+ ATPase [16]. GLUTs are either insulin-dependent (GLUT4) or independent (GLUT1, 2, 3) [17]. Insulin-responsive tissues express low levels of GLUT1 and 3 for basal—stable—glucose uptake and mostly GLUT4 to mediate insulin-induced glucose uptake.

The glucose uptake assay is, together with the evaluation of the insulin-signaling cascade, a way to measure cellular ability to respond to circulating insulin levels. Direct and indirect methods are available to measure insulin-stimulated glucose uptake. Direct methods use radiolabeled (or fluorescently labeled) glucose analogues (^{14}C-labeled 2-deoxyglucose—^{14}C-2DG), which allow the direct measurement of glucose internalization; indirect methods, instead, measure glucose disappearance from the cell culture medium. The choice of methodology includes numerous factors such as sensitivity of detection as well as safety and infrastructural requirements for radiolabel work. Critically though, indirect methods cannot be used for cell-types with significant capacity to release glucose back into the extra-cellular medium, for example hepatocytes. The vast majority of the indirect methods determine glucose using the glucose oxidase–peroxidase–chromogen sequence, where the color intensity is proportional to the quantity of glucose in the sample.

Here we describe both a direct and an indirect approach:

1. ^{14}C-2DG-based glucose uptake [18].

   (a) Differentiate 3T3-L1 or C2C12 cells in 12-well plates as previously described.

   (b) The day before the experiment, exchange FBS from the medium with 1 % BSA and incubate the cells overnight.

   (c) The day of the experiment, stimulate cells for 10 min with Hedgehog modulators as described [8].

   (d) Rapidly wash the wells three times with the reaction buffer (150 mM NaCl, 5 mM KCl, 1.2 mM MgSO$_4$, 1.2 mM NaH$_2$PO$_4$, 10 mM HEPES, and 0.1 % BSA, pH 7.4).

   (e) Incubate the cells in the same reaction buffer supplemented with 10 µM (0.2 µCi/well) of ^{14}C-2DG and let the reaction go for 15 min.

   (f) Stop the reaction by aspirating the reaction mixture and rapidly rinsing each well four times in phosphate buffer saline (PBS) on ice.

   (g) Solubilize cells in 0.5 ml of 0.1 M NaOH and remove 50 µl for the determination of protein concentration.

   (h) Measure glucose uptake by reading the remaining fluid by scintillation.

   (i) For the determination of nonspecific 2DG uptake, incubate sister cultures with 100 µg/ml of Phloretin (*see* **Note 6**).

   (j) Express normalized glucose uptake as pmol 2DG/min/mg of total proteins.

2. Indirect glucose uptake assay.

   (a) Differentiate 3T3-L1 or C2C12 cells in 12-well plates as previously described.

   (b) The day before the experiment, exchange FBS from the medium with 1 % BSA and incubate the cells overnight.

   (c) The day of the experiment, stimulate cells with Hedgehog modulators as described [8].

   (d) At each time point, transfer 2 µl of the cell culture medium into wells of a 96 well plate and add 200 µl of the reaction buffer (225 mM phosphate buffer pH 6.6, 0.3 mM 4-aminoantipyrine, 8.5 mM phenol, 5 mM EDTA, ≥300 U/l peroxidase, ≥10,000 U/l glucose oxidase) (*see* **Note 7**).

   (e) Incubate for 10 min at 37 °C followed by 20 min at 20–25 °C.

   (f) Read absorbance at 505 nm.

(g) Use the reaction buffer alone as a negative control and always use a standard curve prepared with the same culture medium and reaction buffer.

(h) Results are extrapolated from the standard curve and normalized on cell number and/or total DNA content.

3.2.2 Lactate Secretion Assay

Commercial kits are available for the measurement of lactate production and extracellular secretion. For the study of Hh-dependent changes in cellular metabolism we recommend the Lactate PAP assay kit from Biomérieux (Cat#61192). Briefly:

1. Differentiate and treat cells as previously described.

2. At each time point, transfer 2 μl of the cell culture medium into wells of a 96 well plate and add 200 μl of the reaction buffer following the instructions provided with the kit.

3. Incubate the reaction 10 min at room temperature and measure the color intensity at 505 nm.

3.2.3 Oxygen Consumption (OCR) and Extracellular Acidification Rate (ECAR)

Analysis of extracellular flux (XF) in living cells has become a mainstream method to measure cellular bioenergetics. Though indirect, the determination of OCR and ECAR provides detailed information on mitochondrial function.

For the analysis of Hh-dependent changes in cellular metabolism, we used the XF24 and XF96 analysers from Seahorse Bioscience and the general experimental guidelines provided by the company with few modifications.

1. Mitochondrial stress test.

(a) Seed 3T3-L1 preadipocytes and C2C12 myoblasts in the wells of the XF96 Cell Culture Microplates and differentiate as previously described.

(b) The day before the experiment, exchange FBS from the medium with 1 % BSA and incubate the cells overnight (*see* **Note 8**).

(c) On the day of the experiment, pretreat the cells with 200 nM SAG for 6 h.

(d) Change the medium with the Assay Medium (MgSO$_4$ 0.8 mM, CaCl$_2$ 1.8 mM, NaCl 143 mM, KCl 5.4 mM, NaH$_2$PO$_4$ 0.91 mM, L-glutamine 2 mM, Na-pyruvate 2 mM, d-glucose 25 mM). Adjust the medium to pH 7.4 and incubate at least 1 h at 37 °C in a CO$_2$-free incubator (*see* **Note 9**).

(e) Transfer the plate to the XF96 and start the following cycling program: five cycles of 4 min mixture, 2 min wait, and 2 min measurement time.

(f) The final concentrations of the mitochondrial stressors are the following: for adipocytes (Oligomycin 1 μg/ml; FCCP

0.6 μM and Antimycin/Rotenone 3 μM); for myotubes (Oligomycin 2.5 mg/ml; FCCP 0.8 mM and Antimycin/Rotenone 4 μM and 2 μM, respectively).

(g) After the experiment, lyse the cells and use genomic DNA for normalization.

2. Glycolytic stress test.
For the Glycolytic Stress Test, please follow the procedures as previously described with the few following modifications:

(a) Use a 2-port XF96 sensor cartridge because only two compounds are to be injected.

(b) The day of the experiment, change the medium with the assay medium without glucose and incubate at least 1 h at 37 °C in a CO_2-free incubator.

(c) Use the following compound concentrations: d-Glucose 10 mM and 2-DG 10 mM.

3. Phenotype microarrays.
The Phenotype Microarray (PM—http://www.biolog.com) is an "in-cell" assay that allows the determination of metabolic and chemical sensitivity phenotypes of mammalian cells. The assay consists of 96 well plates coated with metabolites or chemicals. A cell suspension is added to each well and the use of individual substrates is measured by quantifying the reduction of a colorimetric redox dye as result of the generation of energy-rich NADH by the cells. The assay is briefly carried out as follows:

(a) Grow and differentiate 3T3-L1 and C2C12 cells as previously described.

(b) Inoculate 50 μl of cell suspension (in IF-M1 medium provided with the PM plates) per well of PM-M1—PM-M4 (carbon and nitrogen energy sources) and incubate for 16 h at 37 °C 5 % CO_2.

(c) The day after, add 10 μl/well of the BIOLOG Redox Dye MA and quantitate substrates use at 590 and 750 nm (*see* **Note 10**).

3.3 Characterization of Molecular Mechanisms

3.3.1 Calcium Signal Measurement

1. Plate, grow and differentiate 3T3-L1 and C2C12 cells in black wall/clear bottom 96-well plates as already described.

2. The day before the experiment, exchange FBS from the medium with 1 % BSA and incubate the cells overnight.

3. On the day of the experiment, load cells with the calcium dye FluoForte AM (Enzo LifeScience) according to the manufacturer's instructions.

4. Measure calcium flux by detecting fluorescence at 488/530 nm (excitation/emission) before and after injection of the Hh modulators at the indicated concentrations.

5. Use Ionomycin (1 µM) and EGTA (1 µM) as a positive control and extracellular calcium chelator, respectively.

3.3.2 Real-Time Measurement of Ampk Activity

Ampk activity can be measured by using the Foerster Resonance Energy Transfer (FRET)-based Ampk biosensor (AMPK-AR) generated by Tsou et al. [19]. Briefly:

1. Generate HEK293 cells stably expressing the AMPK-AR construct and resuspend them in PBS at a concentration of 10^5 cells/ml.

2. Prepare a black wall/clear bottom 96-well plate with the Hh modulators pipetted in single wells at the indicated concentration.

3. Pre-load the cell suspension into the injection system of the reader (e.g., BioTek Synergy 4) and inject them at 10^4 cells/well.

4. Detect FRET signals at 420/50 and 528/20 nm as excitation and emission settings.

5. Acquire a baseline for each well before the injection of cells to be subtracted afterwards.

3.4 Mouse Studies

All our mouse studies are performed on C57BL/6J maintained on a standard rodent chow diet and a 12 h light–12 h dark cycle.

For the vast majority of the in vivo analysis of Hedgehog-dependent metabolic effects, we used cyclopamine as tool compound. Cyclopamine is the prototype for competitive inhibitors of the canonical Hedgehog signaling pathway, and therefore it has been widely studied in vivo. Data on solubility, PK/PD profile and toxicity are available [20], and represent a precious starting point for such in vivo work [8]. We describe here three different mouse experiments aimed at studying the role of Hh partial agonism as a potential insulin sensitizer (Subheading 3.4.1); as an insulin mimetic (Subheading 3.4.2); and as a tissue-specific modulator of glucose uptake (Subheading 3.4.3).

3.4.1 Oral Glucose Tolerance Test (OGTT)

1. The night before the experiment, fast 8 weeks-old wild-type C57BL/6J mice for 12 h.

2. On the day of the experiment, pre-treat the animals with a single dose of cyclopamine (10 mg/kg dissolved in 2-hydroxypropyl-beta-cyclodextrin in PBS (or HBS) as described [8, 20]) or vehicle alone.

3. After 4 h (*see* **Note 11**), start the OGTT by sampling blood from the tail vein of each animal to determine basal insulin and glucose levels.

4. Administer 2 g/kg glucose by oral gavage and sample blood every 15 min (from 0 to 60) to measure glucose and insulin levels.

5. For the determination of insulin levels we used the Ultrasensitive Mouse Elisa Insulin Detection Kit (Mercodia) according to the manufacturer's instructions.

3.4.2 Cyclopamine Tolerance Test

In order to prove the insulin-independency of the cyclopamine-induced glucose uptake and its therapeutic relevance, wild-type C57BL/6J mice were made diabetic by single high-dose Streptozotocin (STZ) treatment. STZ is an alkylating agent of the nitrosurea class, whose toxicity is quite specific for pancreatic β-cells [21, 22]. For this reason, STZ has been approved by the Food and Drug Administration (FDA) for the treatment of pancreatic cancers that are not eligible for surgery and used as a tool compound to generate mouse models of insulitis (multiple low-dose) and insulin-dependent type 1 diabetes (single high-dose).

1. Eight weeks-old wild-type C57BL/6J mice are injected intraperitoneally (i.p.) with a single STZ dose (200 mg/kg of body weight) to destroy pancreatic β-cells (*see* **Note 12**).

2. 72 h later, separate the STZ-treated animals into two groups and inject them i.p. with the vehicle or cyclopamine (10 mg/kg, see before how to prepare cyclopamine for mouse studies).

3. Record blood glucose by tail vein bleeding at the beginning and every 15 min during the experiment.

4. To investigate Hh-dependent induction of non-shivering thermogenesis and brown adipose tissue activation, also record core body temperature by using a rectal probe.

3.4.3 Glycemic-Insulinemic Clamp Studies

For the clamp studies, refer to the general procedure as published by Knauf et al. and Pospisilik et al. [13, 23]. In order to study Hh-dependent tissue-specific glucose uptake and the insulin independence of the phenotype, we modified the published protocol as follows:

1. After having allowed the mice to recover from surgery for 5 days, fast them 16 h before the experiment.

2. On the day of the experiment, infuse somatostatin (SST14) at a rate of 1 μg/kg/min and maintain hyperglycemia (12.5 mM) by variable infusion of a 10 % glucose solution.

3. Maintain the steady state for 90 min before injection of cyclopamine (2/3 bolus 1/3 infusion; 10 mg/kg) or HBS vehicle.

4. Continue measuring blood glucose every 10 min for 1 h.

5. Measure tissue-specific glucose uptake as described [13, 23].

4   Notes

1. Sufficient differentiation is achieved without adding TGZ to the induction and differentiation cocktails [24]. To measure Hh-dependent metabolic effects, though, TGZ is absolutely essential. Adipocytes differentiated without TGZ do not respond to Hh. TGZ treatment stimulates mitochondrial biogenesis via Peroxisome Proliferator-Activated Receptor Gamma (PPARG) [25], and induces the expression of brown fat-specific genes [25]. Therefore, TGZ treatment increases the overall cellular metabolic capacity, upon which the Smo-Ampk noncanonical Hh signaling relies to rewire metabolism, as highlighted by Teperino et al. [8].

2. When replacing the medium it is important to leave 10 % of the old medium in the culture dish. Differentiating adipocytes secrete autocrine factors, which further sustain adipocyte differentiation [26].

3. We learnt from our experience that adipocytes at odd days of differentiation are more sensitive to metabolic stimuli, most likely because medium exchange re-sets signaling cascades and prevent negative feedback loops.

4. The concentrations of the Hh modulators generally used for the described metabolic studies are as follow: Smoothened Agonist (SAG—200 nM in DMSO); Cyclopamine (Cyc—100 nM in DMSO); GDC-0449 (10—100 nM in DMSO); and rSHH (0.25 μg/ml).

5. It is equally important with C2C12 cultures, during the maintenance period, to avoid confluency to preserve the full differentiation potential.

6. Phloretin is a glucose analogue and potent inhibitor of glucose uptake [27].

7. Given that the reaction buffer is linear up to 4 g/l, it is recommended to dilute samples accordingly.

8. The day before the experiment it is important to calibrate the XF96 4-port Sensor Cartridge with the provided Calibrant (200 μl/well) and incubate at 37 °C in a CO_2-free incubator.

9. This incubation period is essential for the stabilization of the baseline. Adjust the incubation time according to experimental results.

10. As indicated in the BIOLOG instructions book, subtract the 790 nm reading (as background reading).

11. As shown by Lipinski et al. [20], i.p. injection of cyclopamine at 10 mg/kg achieves, within 4 h, the desired serum concentration of 100 nM, which is what we used for all the in vitro assays.

12. As mentioned in the text, STZ treatment induces acute insulitis, a burst of massive insulin release associated with β-cell death and consequently hypoglycemia. For this reason, it is crucial, especially during the first hours of STZ treatment, to follow blood glucose concentration of the animals in order to prevent any lethal hypoglycemia. After the first days, insulitis gives rise to overt type 1 diabetes.

Acknowledgements

JAP acknowledges generous support from the DFG, ERC, EU-FP7, BMBF (DEEP), and the MPG.

References

1. Pires-daSilva A, Sommer RJ (2003) The evolution of signalling pathways in animal development. Nat Rev Genet 4(1):39–49

2. Aberger F, Ruiz I, Altaba A (2014) Context-dependent signal integration by the GLI code: the oncogenic load, pathways, modifiers and implications for cancer therapy. Semin Cell Dev Biol 33:93–104

3. Briscoe J, Thérond PP (2013) The mechanisms of hedgehog signalling and its roles in development and disease. Nat Rev Mol Cell Biol 14(7):416–429

4. Teperino R, Aberger F, Esterbauer H, Riobo N, Pospisilik JA (2014) Canonical and non-canonical hedgehog signalling and the control of metabolism. Semin Cell Dev Biol 33:81–92

5. Mukhopadhyay S, Rohatgi R (2014) G-protein-coupled receptors, hedgehog signaling and primary cilia. Semin Cell Dev Biol 33:63–72

6. Ingham PW, McMahon AP (2001) Hedgehog signaling in animal development: paradigms and principles. Genes Dev 15(23):3059–3087

7. Brennan D, Chen X, Cheng L, Mahoney M, Riobo NA (2012) Noncanonical hedgehog signaling. Vitam Horm 88:55–72

8. Teperino R, Amann S, Bayer M, McGee SL, Loipetzberger A, Connor T et al (2012) Hedgehog partial agonism drives Warburg-like metabolism in muscle and brown fat. Cell 151(2):414–426

9. Amakye D, Jagani Z, Dorsch M (2013) Unraveling the therapeutic potential of the hedgehog pathway in cancer. Nat Med 19(11):1410–1422

10. Mullor JL, Sánchez P, Ruiz i Altaba A (2002) Pathways and consequences: Hedgehog signaling in human disease. Trends Cell Biol 12(12):562–569

11. Collins S, Martin TL, Surwit RS, Robidoux J (2004) Genetic vulnerability to diet-induced obesity in the C57BL/6J mouse: physiological and molecular characteristics. Physiol Behav 81(2):243–248

12. Cantó C, Gerhart-Hines Z, Feige JN, Lagouge M, Noriega L, Milne JC et al (2009) AMPK regulates energy expenditure by modulating NAD+ metabolism and SIRT1 activity. Nature 458(7241):1056–1060

13. Pospisilik JA, Schramek D, Schnidar H, Cronin SJ, Nehme NT, Zhang X et al (2010) Drosophila genome-wide obesity screen reveals hedgehog as a determinant of brown versus white adipose cell fate. Cell 140(1):148–160

14. Gaster M, Kristensen SR, Beck-Nielsen H, Schröder HD (2001) A cellular model system of differentiated human myotubes. APMIS 109(11):35–44

15. Gaster M, Schröder HD, Handberg A, Beck-Nielsen H (2001) The basal kinetic parameters of glycogen synthase in human myotube cultures are not affected by chronic high insulin exposure. Biochim Biophys Acta 1537(3):211–221

16. DeFronzo RA, Davidson JA, Del Prato S (2012) The role of the kidneys in glucose homeostasis: a new path towards normalizing glycaemia. Diabetes Obes Metab 14(1):5–14

17. Mueckler M, Thorens B (2013) The SLC2 (GLUT) family of membrane transporters. Mol Aspects Med 34(2–3):121–138

18. Gaster M, Petersen I, Højlund K, Poulsen P, Beck-Nielsen H (2002) The diabetic phenotype is conserved in myotubes established from diabetic subjects: evidence for primary defects in glucose transport and glycogen synthase activity. Diabetes 51(4):921–927

19. Tsou P, Zheng B, Hsu CH, Sasaki AT, Cantley LC (2011) A fluorescent reporter of AMPK activity and cellular energy stress. Cell Metab 13(4):476–486

20. Lipinski RJ, Hutson PR, Hannam PW, Nydza RJ, Washington IM, Moore RW et al (2008) Dose- and route-dependent teratogenicity, toxicity, and pharmacokinetic profiles of the hedgehog signaling antagonist cyclopamine in the mouse. Toxicol Sci 104(1):189–197

21. Sakata N, Yoshimatsu G, Tsuchiya H, Egawa S, Unno M (2012) Animal models of diabetes mellitus for islet transplantation. Exp Diabetes Res 2012:256707

22. Damasceno DC, Netto AO, Iessi IL, Gallego FQ, Corvino SB, Dallaqua B et al (2014) Streptozotocin-Induced diabetes models: pathophysiological mechanisms and fetal outcomes. Biomed Res Int 2014:819065

23. Knauf C, Cani PD, Perrin C, Iglesias MA, Maury JF, Bernard E et al (2005) Brain glucagon-like peptide-1 increases insulin secretion and muscle insulin resistance to favor hepatic glycogen storage. J Clin Invest 115(12):3554–3563

24. Zebisch K, Voigt V, Wabitsch M, Brandsch M (2012) Protocol for effective differentiation of 3T3-L1 cells to adipocytes. Anal Biochem 425(1):88–90

25. Rong JX, Klein JL, Qiu Y, Xie M, Johnson JH, Waters KM et al (2011) Rosiglitazone induces mitochondrial biogenesis in differentiated murine 3T3-L1 and C3H/10T1/2 adipocytes. PPAR Res 2011:179454

26. Hemmingsen M, Vedel S, Skafte-Pedersen P, Sabourin D, Collas P, Bruus H, Dufva M (2013) The role of paracrine and autocrine signaling in the early phase of adipogenic differentiation of adipose-derived stem cells. PLoS One 8(5):e63638

27. Frerichs H, Ball EG (1964) Studies on the metabolism of adipose tissue. XVI. Inhibition by phlorizin and phloretin of the insulin-stimulated uptake of glucose. Biochemistry 3:981–985

INDEX

Natalia A. Riobo (ed.), *Hedgehog Signaling Protocols*, Methods in Molecular Biology, vol. 1322,
DOI 10.1007/978-1-4939-2772-2, © Springer Science+Business Media New York 2015